Lecture Notes in Control and Information Sciences

Volume 454

T0212163

For further volumes:
http://www.springer.com/series/642

About this Series

This series aims to report new developments in the fields of control and information sciences - quickly, informally and at a high level. The type of material considered for publication includes:

1. Preliminary drafts of monographs and advanced textbooks
2. Lectures on a new field, or presenting a new angle on a classical field
3. Research reports
4. Reports of meetings, provided they are   a) of exceptional interest and
b) devoted to a specific topic. The timeliness of subject material is very important.

Grzegorz Mzyk

Combined Parametric-Nonparametric Identification of Block-Oriented Systems

 Springer

Grzegorz Mzyk
Institute of Computer Engineering, Control
   and Robotics
Wrocław University of Technology
Wrocław
Poland

ISSN 0170-8643   ISSN 1610-7411   (electronic)
ISBN 978-3-319-03595-6   ISBN 978-3-319-03596-3   (eBook)
DOI 10.1007/978-3-319-03596-3
Springer Cham Heidelberg New York Dordrecht London

Library of Congress Control Number: 2013954682

Printed on acid-free paper

Springer is part of Springer Science+Business Media (www.springer.com)

To my wife Jana, and my children – Adam and Hanna

Preface

Should we trust the prior knowledge of the system before processing observations? It is one of fundamental questions in system identification. Correct parametric assumptions imposed on the form of input output relation obviously give chances for obtaining relatively accurate model, using only several measurements. On the other hand, if the taken assumptions were false we would risk the non-zero approximation error, which could not be avoided even for large number of measurements.

The main purpose of this book is to provide a set of flexible identification tools, which allow to work under partial or uncertain prior knowledge, support making decision of the model structure and improve accuracy of the model for small number of measurements.

The presented ideas combine methods of two different philosophies – traditional (parametric) and recent (nonparametric), trying to take advantages of both of them. The term 'parametric' means that the assumed class of model consists of *finite* and *known* number of parameters and we look for optimal parameters in the finite dimensional space, using observations (e.g. by the least squares method, proposed by Gauss in 1809). Nonparametric estimates ([50], [100]) use only observations to build the model, and the model becomes more and more complex, when the number of data increases.

Combined identification schemes, proposed in this book, can be divided into two groups:

- *mixed parametric-nonparametric*, in which, nonparametric algorithm is first applied for estimation of interaction signals in complex system, or generation of instrumental variables, and then, the results are used (plugged-in) in parameter stage; these methods are recommended when the prior knowledge of the system is partial,
- *semiparametric*, in which, parametric model is firstly computed in traditional way, and the nonparametric correction is added when the number of measurements grows large; these methods allow to reduce the error for small number of data.

In Chapter 1 we briefly present foundations of the identification of single-element objects (linear dynamic or static nonlinear). As regards the parameter estimation, we focus on the least squares and instrumental variables methods, which play the key role in the book. Also nonparametric techniques, i.e., kernel regression and orthogonal series expansion are shortly reminded. We start our considerations in Chapter 2, from the Hammerstein system (i.e. the simplest N-L tandem connection), with FIR dynamics, linear-in-the-parameters static characteristics, and i.i.d. excitation. Next, the assumptions are generalized for the IIR subsystem, hard nonlinearity and correlated input case. Similar analysis is made for a Wiener system (reverse connection) in Chapter 3. Chapters 4-6 are devoted to identification of systems with more complex structures, i.e. Wiener-Hammerstein (L-N-L sandwich) system, the additive NARMAX system, and the system with general structure of interconnections. The model selection problem is considered in Chapter 7, and the time-varying block-oriented systems are identified in Chapter 8. Proofs of theorems, and technical lemmas can be found in two Appendices at the end.

Some results of the book were presented at the following conferences: *IFAC World Congress 2008*, Seoul, Korea, *SYSID 2009*, Saint-Malo, France, *ALCOSP 2010*, Antalya, Turkey, *SYSID 2012*, Brussels, Belgium, *ICINCO 2012*, Rome, Italy, *ECT 2012*, Kaunas, Lithania, and *ICCC 2012*, Kosice, Slovakia.

I would like to thank my co-workers from the Department of Control and Optimization, Wrocław University of Technology, in which I have pleasure to work. Special thanks go to prof. Zygmunt Hasiewicz and prof. Włodzimierz Greblicki, my teachers in control theory and system identification, for arranging comfortable conditions, which helped me to write this book. I would also like to thank colleagues from the room – Dr. Przemysław Śliwiński and Dr. Paweł Wachel for scientific atmosphere, discussions, and numerous suggestions.

This work was supported by Polish National Science Centre (Grant No. N N514 700640).

Abstract

This book considers a problem of block-oriented nonlinear dynamic system identification in the presence of random disturbances. The class of systems includes various interconnections of linear dynamic blocks and static nonlinear elements, e.g., Hammerstein system, Wiener system, Wiener-Hammerstein ("sandwich") system and additive NARMAX systems with feedback. Interconnecting signals are not accessible for measurement. The combined parametric-nonparametric algorithms, proposed in the book, can be selected dependently on the prior knowledge of the system and signals. Most of them is based on the decomposition of the complex system identification task into simpler local subproblems by using nonparametric (kernel or orthogonal) regression estimation. In the parametric stage, the generalized least squares or the instrumental variables technique is commonly applied to cope with correlated excitations. Limit properties of the algorithms have been shown analytically and illustrated in simple experiments.

Contents

Chapter 1
Introduction

1.1 Parametric Identification of Single-Element Systems

In this section we recall the most important facts concerning the least squares (LS) and instrumental variables (IV) methods (see e.g. [128] and [143]). Firstly, we consider the multiple input single output static element (MISO), and next, the single input single output (SISO) linear dynamic object, with finite impulse response (FIR) and with infinite impulse response (IIR). Finally, the concept of the instrumental variables technique is explained, to reduce the bias of least squares estimate for correlated disturbances. In the whole book we deal with the time-domain methods. We refer the readers interested in frequency-domain modeling to [120], [109], and [76].

1.1.1 Least Squares Approach

Let us consider the MISO static system described by the equation

$$y = F^*(x, z) = F(x, z, a^*),$$

where the input $x = \left(x^{(1)}, x^{(2)}, ..., x^{(s)}\right)^T$ is accessible for a direct measurement, $a^* = (a_1^*, a_2^*, ..., a_s^*)^T$ is an unknown vector of true parameters, $F()$ is the class of function of known form, and z is a random disturbance. We look for the true system description $F^*()$ assuming that it belongs to the class $F()$ (prior knowledge), i.e.,

$$F^* \in \{F(x, z, a); \ a \in R^s\} \tag{1.1}$$

and that the solution is unique with respect to parameters (identifiability condition), i.e.,

$$F^*(x, z) = F(x, z, a) \Longrightarrow a = a^*,$$

G. Mzyk, *Combined Parametric-Nonparametric Identification*
of Block-Oriented Systems, Lecture Notes in Control and Information Sciences 454,
DOI: 10.1007/978-3-319-03596-3_1, © Springer International Publishing Switzerland 2014

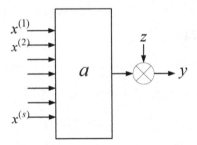

Fig. 1.1 The *MISO* static system

using the set of N input-output observations (learning sequence)

$$\{x_k, y_k\}_{k=1}^N. \tag{1.2}$$

The goal is to compute the estimate $\widehat{a}_N$ of unknown a^*, using these two kinds of knowledge (prior knowledge (1.1) and measurements (1.2)).

Assume that the noise is additive (see Fig. 1.1)

$$y = F(x, a^*) + z,$$

where $\{z_k\}_{k=1}^N$ is i.i.d. zero-mean random sequence with finite variance, and independent of x, i.e.,

$$Ez = 0, \quad \mathrm{var}z < \infty, \quad z, x - \text{independent}. \tag{1.3}$$

The aim is to minimize the model quality index of the form

$$Q(a) = E\{y - F(x, a)\}^2 \to \min_a.$$

Taking into account that

$$
\begin{aligned}
Q(a) &= E\{F(x, a^*) + z - F(x, a)\}^2 = \\
&= E\{[F(x, a^*) - F(x, a)]^2 + 2[F(x, a^*) - F(x, a)]z + z^2\} = \\
&= E\{F(x, a^*) - F(x, a)\}^2 + \mathrm{var}z,
\end{aligned}
\tag{1.4}
$$

we conclude that

$$Q(a) \ge \mathrm{var}z,$$

and

$$a^* = \arg\min_a Q(a).$$

In practice, $Q(a)$ cannot be computed (unknown distributions of excitations), and the expectation in (1.4) is replaced by the sample mean value [Gauss, 1809]

$$\widehat{Q}_N(a) = \frac{1}{N} \sum_{k=1}^{N} \{y_k - F(x_k, a)\}^2 \rightarrow \min_a. \qquad (1.5)$$

The components $\{y_k - F(x_k, a)\}^2$ in the sum (1.5) are i.i.d., thus, the strong law of large numbers works, and it holds that (see Appendix B.4)

$$\widehat{Q}_N(a) \xrightarrow{p.1} Q(a), \text{ as } N \rightarrow \infty.$$

Moreover, if

$$\sup_a(\widehat{Q}_N(a) - Q(a)) \xrightarrow{p.1} 0,$$

then also

$$\arg\min_a \widehat{Q}_N(a) \xrightarrow{p.1} \arg\min_a Q(a) = a^*.$$

MISO Linear Static Element

For the linear function $F()$

$$y = a^{*T}x + z = x^T a^* + z,$$
$$y_k = x_k^T a^* + z_k, \qquad k = 1, 2, ..., N,$$

the whole experiment can be described in the compact matrix-vector form (measurement equation)

$$Y_N = X_N a^* + Z_N, \qquad (1.6)$$

where

$$X_N = \begin{bmatrix} x_1^T \\ x_2^T \\ \vdots \\ x_N^T \end{bmatrix} = \begin{bmatrix} x_1^{(1)} & x_1^{(2)} & .. & x_1^{(s)} \\ x_2^{(1)} & x_2^{(2)} & .. & x_2^{(s)} \\ .. & .. & .. & .. \\ x_N^{(1)} & x_N^{(2)} & .. & x_N^{(s)} \end{bmatrix},$$

$$Y_N = \begin{bmatrix} y_1 \\ y_2 \\ \vdots \\ y_N \end{bmatrix}, \text{ and } Z_N = \begin{bmatrix} z_1 \\ z_2 \\ \vdots \\ z_N \end{bmatrix}.$$

We consider a class of linear models

$$\overline{Y}_N = X_N a$$

and minimize the following empirical criterion of the model quality

$$\overline{Q}_N(a) = \sum_{k=1}^{N}\{y_k - x_k^T a\}^2 = (Y_N - X_N a)^T(Y_N - X_N a) =$$
$$= \|Y_N - X_N \mathbf{a}\|_2^2 = (Y_N^T - a^T X_N^T)(Y_N - X_N a) =$$
$$= Y_N^T Y_N - a^T X_N^T Y_N - Y_N^T X_N a + a^2 X_N^T X_N \to \min_a,$$

where $\|\cdot\|_2$ denotes Euclidean vector norm. The gradient

$$\nabla_a \overline{Q}_N(a) = -2X_N^T Y_N + 2X_N^T X_N a$$

is zero for

$$X_N^T X_N a = X_N^T Y_N, \tag{1.7}$$

where $\dim X_N^T X_N = s \times s$, $\dim X_N^T Y_N = s \times 1$, and $\dim a = s \times 1$. The solution of (1.7) with respect to a exists if $\mathrm{lincol} X_N^T X_N = \mathrm{lincol} X_N^T$ and is unique if $X_N^T X_N$ is of full rank, i.e. $\mathrm{rank} X_N^T X_N = s$ and equivalently $\det X_N^T X_N \neq 0$. Obviously, the number of measurements must not be less than number of parameters (necessary condition). Moreover at least s of N measurements must be linearly independent (sufficient condition). If $\mathrm{rank} X_N^T X_N < s$, then the solution is not unique. Since the elements in X_N and Y_N are random, the task (1.7) can be badly conditioned numerically and special numerical procedures can be found in the literature, e.g. LU decomposition, singular value decomposition, etc. Also recursive version

$$\hat{a}_N = \hat{a}_{N-1} + P_N x_N (y_N - x_N^T \hat{a}_{N-1}),$$
$$P_N = P_{N-1} - \frac{P_{N-1} x_N x_N^T P_{N-1}}{1 + x_N^T P_{N-1} x_N},$$
$$P_0 = diag[10^2 \div 10^5], \ a_0 = 0,$$

can be applied (see Appendix A.1).

Normal Equation

The equation (1.7) can be rewritten in the form

$$X_N^T(Y_N - X_N a) = 0.$$

It means that, the model is optimal, if the difference between system outputs Y_N and model outputs $\overline{Y}_N = X_N a$ is orthogonal to the columns of the matrix X_N^T (see Fig. 1.2)

$$X_N^T = [col_1 x, col_2 x, ..., col_s x].$$

Hence, (1.7) is often called the 'normal equation' and its solution

$$\hat{a}_N = \left(X_N^T X_N\right)^{-1} X_N^T Y_N \tag{1.8}$$

is called the 'least squares estimate' of a.

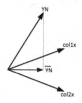

Fig. 1.2 Orthogonal projection

Limit Properties ($N \to \infty$)

Multiplying the equation $y = x^T a^* + z$ by x we obtain that $xy = xx^T a^* + xz$ and obviously $Exy = Exx^T a^* + Exz$. Since x and z are independent, and $Ez = 0$ we have

$$a^* = (Exx^T)^{-1} Exy, \tag{1.9}$$

provided that Exx^T is not singular (the input must be 'rich' enough). Let us notice that

$$X_N^T X_N = \sum_{k=1}^N x_k x_k^T, \qquad X_N^T Y_N = \sum_{k=1}^N x_k y_k,$$

and rewrite the *LS* estimate (1.8) in the form

$$\widehat{a}_N = \left(\frac{1}{N} \sum_{k=1}^N x_k x_k^T \right)^{-1} \left(\frac{1}{N} \sum_{k=1}^N x_k y_k \right).$$

where the components $\{x_k x_k^T\}$ and $\{x_k y_k\}$ are i.i.d. sequences. Hence,

$$\frac{1}{N} \sum_{k=1}^N x_k x_k^T \xrightarrow{p.1} Exx^T, \text{ as } N \to \infty,$$

$$\frac{1}{N} \sum_{k=1}^N x_k y_k \xrightarrow{p.1} Exy, \text{ as } N \to \infty, \tag{1.10}$$

and by continuity of the inversion operation $()^{-1}$, also

$$\left(\frac{1}{N} \sum_{k=1}^N x_k x_k^T \right)^{-1} \xrightarrow{p.1} (Exx^T)^{-1}, \text{ as } N \to \infty. \tag{1.11}$$

Consequently, it holds that (see (1.9), (A.16), and (A.17))

$$\widehat{a}_N \xrightarrow{p.1} a^*, \text{ as } N \to \infty.$$

Bias and Covariance Matrix ($N < \infty$)

Introducing the symbol

$$L_N = \left(X_N^T X_N\right)^{-1} X_N^T$$

we see that the estimate

$$\widehat{a}_N = L_N Y_N$$

is linear (with respect to outputs Y_N). Since $Y_N = X_N a^* + Z_N$, we obtain that

$$E\widehat{a}_N = E\left\{L_N(X_N a^* + Z_N)\right\} = a^* + EL_N \cdot EZ_N = a^*,$$

so the LS estimate is unbiased for each N. Moreover, as it was shown in Appendix A.2, its covariance matrix

$$cov(\widehat{a}_N) = \sigma_z^2 \left(X_N^T X_N\right)^{-1}$$

is minimal in the class of linear unbiased estimates (LUE).

SISO Linear Dynamic Object

In this point we show the analogies between formal descriptions of *MISO static system* and *SISO dynamic object*. We first consider the moving average (*MA*) model with the finite impulse response (*FIR*) and show that all properties of the least squares (*LS*) estimate remain unchanged. Next, we assume the *ARMA* object, with the infinite impulse response (*IIR*) and show the reason of the bias. Finally, the idea of instrumental variables (*IV*) method, which allows to reduce the bias, is presented.

Finite Memory Dynamics (FIR)

Let the discrete-time linear dynamic object be described by the following difference equation

$$v_k = b_0 u_k + \ldots + b_s u_{k-s} \tag{1.12}$$

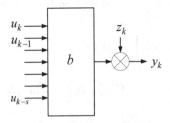

Fig. 1.3 The *MA* linear dynamic object

and only noise-corrupted output y_k can be observed, i.e.,

$$y_k = v_k + \varepsilon_k, \qquad (1.13)$$

where $E\varepsilon_k = 0$, $\mathrm{var}\varepsilon_k < \infty$, $\{\varepsilon_k\}$ is i.i.d. and independent of $\{u_k\}$. The object can be plotted analogously as the *MISO* static system with the inputs $u_k, u_{k-1}, ..., u_{k-s}$ (see Fig. 1.3). From the parametric identification point of view we need the direct relation between observed signals $\{y_k\}$ and $\{u_k\}$. From (1.12) and (1.13) we simply get

$$y_k = b_0 u_k + ... + b_s u_{k-s} + z_k \qquad (1.14)$$

where the sequence $\{z_k\}$ appearing in (1.14) and in Fig. 1.3 is obviously equivalent to the noise $\{\varepsilon_k\}$, i.e.,

$$z_k = \varepsilon_k.$$

It leads to the following conclusion.

Conclusion 1.1. *The least squares estimate of the parameter vector* $b = (b_0, b_1, ..., b_s)^T$ *of the form*

$$\widehat{b}_N = \left(\Phi_N^T \Phi_N\right)^{-1} \Phi_N^T Y_N, \qquad (1.15)$$

where $Y_N = (y_1, y_2, ..., y_N)^T$ *and*

$$\Phi_N = \begin{bmatrix} \phi_1^T \\ \phi_2^T \\ \vdots \\ \phi_N^T \end{bmatrix} = \begin{bmatrix} u_1 & u_0 & .. & u_{1-s} \\ u_2 & u_1 & .. & u_{2-s} \\ \vdots & \vdots & \vdots\vdots & \vdots \\ u_N & u_{N-1} & .. & u_{N-s} \end{bmatrix}, \qquad (1.16)$$

has the same properties as (1.8), i.e., $E\widehat{b}_N = b$, *and* $\widehat{b}_N \to b$ *with probability 1, as* $N \to \infty$.

Infinite Memory Dynamics (IIR)

For the *ARMA* object with the output noise ε_k

$$v_k = b_0 u_k + ... + b_s u_{k-s} + a_1 v_{k-1} + ... + a_p v_{k-p}, \qquad (1.17)$$
$$y_k = v_k + \varepsilon_k,$$

we obtain the following relation between observed y_k and the input u_k

$$y_k = b_0 u_k + ... + b_s u_{k-s} + a_1 y_{k-1} + ... + a_p y_{k-p} + z_k, \qquad (1.18)$$

in which, the sequence $\{z_k\}$(see Fig. 1.4) is not equivalent to the noise $\{\varepsilon_k\}$, but it is colored version of $\{\varepsilon_k\}$

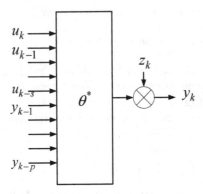

Fig. 1.4 The $ARMA$ linear dynamic object

$$z_k = \varepsilon_k - a_1\varepsilon_{k-1} - \ldots - a_p\varepsilon_{k-p}.$$

Hence, $\{z_k\}$ is correlated process even if $\{\varepsilon_k\}$ is i.i.d., and consequently, the random variable z_k is correlated with $y_{k-1}, y_{k-2}, \ldots, y_{k-p}$ playing the role of inputs in Fig. 1.4. It leads to the following conclusion.

Conclusion 1.2. *The least squares estimate of the parameter vector* $\theta = (b_0, b_1, \ldots, b_s, a_1, a_2, \ldots, a_p)^T$, *of the form*

$$\widehat{\theta}_N = \left(\varPhi_N^T \varPhi_N\right)^{-1} \varPhi_N^T Y_N, \tag{1.19}$$

where $Y_N = (y_1, y_2, \ldots, y_N)^T$ *and*

$$\varPhi_N = \begin{bmatrix} \phi_1^T \\ \phi_2^T \\ \vdots \\ \phi_N^T \end{bmatrix} = \begin{bmatrix} u_1 & u_0 & \cdots & u_{1-s} & y_0 & y_{-1} & \cdots & y_{1-p} \\ u_2 & u_1 & \cdots & u_{2-s} & y_1 & y_0 & \cdots & y_{2-p} \\ \vdots & \vdots & & \vdots & \vdots & \vdots & & \vdots \\ u_N & u_{N-1} & \cdots & u_{N-s} & y_{N-1} & y_{N-2} & \cdots & y_{N-p} \end{bmatrix}, \tag{1.20}$$

is generally biased, $E\widehat{b}_N \neq b$, *even asymptotically, i.e.* $\lim_{N\to\infty} E\widehat{b}_N \neq b$.

One of the method, which allows to reduce the bias of (1.19) is the Gauss-Markov estimate

$$\widehat{\theta}_N^{(GM)} = \left(\varPhi_N^T C_N \varPhi_N\right)^{-1} \varPhi_N^T C_N Y_N,$$

in which $C_N = \mathrm{cov}(Z_N)$. Nevertheless, since the covariance matrix C_N of the 'noise' Z_N is unknown, the method is difficult to apply. In the next point we present more universal approach, based on generation of additional matrix of instrumental variables.

1.1.2 Instrumental Variables Method

The instrumental variables (IV) estimate of the vector θ has the form ([154], [127], [102])

$$\widehat{\theta}_N^{(\mathrm{IV})} = \left(\Psi_N^T \Phi_N\right)^{-1} \Psi_N^T Y_N,$$

where $\Psi_N = (\psi_1, \psi_2, ..., \psi_{s+p+1})$ is additional matrix, of the same dimensions as Φ_N, i.e.,

$$\dim \Psi_N = \dim \Phi_N,$$

which fulfills the following two postulates with probability 1, as $N \to \infty$

$$\frac{1}{N} \Psi_N^T \Phi_N = \sum_{k=1}^{N} \psi_k \phi_k^T \to E\psi_k \phi_k^T, \text{ where } \det E\psi_k \phi_k^T \neq 0, \qquad (1.21)$$

and

$$\frac{1}{N} \Psi_N^T Z_N = \sum_{k=1}^{N} \psi_k z_k \to 0. \qquad (1.22)$$

The postulate (1.21) assures invertibility of $\Psi_N^T \Phi_N$ and implies correlation between elements of Ψ_N (instruments) and inputs u_k. The condition (1.22) means that the instruments must be asymptotically uncorrelated with the 'noise' z_k. Since z_k is correlated with outputs $y_{k-1}, y_{k-2}, ..., y_{k-p}$, the choice $\Psi_N = \Phi_N$ is obviously excluded. The most popular methods of generation of Ψ_N are based on the linear transformation of the process $\{u_k\}$. Also the least squares can be used as a pilot estimate, to replace the system outputs in Φ_N, with the model outputs, i.e.,

$$\Psi_N = \begin{bmatrix} \psi_1^T \\ \psi_2^T \\ \vdots \\ \psi_N^T \end{bmatrix} = \begin{bmatrix} u_1 & u_0 & \cdots & u_{1-s} & \overline{y}_0 & \overline{y}_{-1} & \cdots & \overline{y}_{1-p} \\ u_2 & u_1 & \cdots & u_{2-s} & \overline{y}_1 & \overline{y}_0 & \cdots & \overline{y}_{2-p} \\ \vdots & \vdots & \vdots & \vdots & \vdots & \vdots & \vdots & \vdots \\ u_N & u_{N-1} & \cdots & u_{N-s} & \overline{y}_{N-1} & \overline{y}_{N-2} & \cdots & \overline{y}_{N-p} \end{bmatrix},$$

where $\overline{y}_k$ denotes the output of the least squares model excited by the true process $\{u_k\}$.

Directly from the Slutzky theorem we obtain the following conclusion.

Conclusion 1.3. *If the instrumental variables matrix fulfills conditions (1.21) and (1.22), then it holds that*

$$\widehat{\theta}_N^{(\mathrm{IV})} \to \theta$$

with probability 1, as $N \to \infty$.

1.2 Nonparametric Methods

In this section we shortly describe methods, which does not assume representation of the system in finite dimensional parameter space. The assumptions
imposed on the nonlinear characteristic or linear dynamics are very poor and
have only general character, e.g., we assume continuity of unknown function or stability of linear dynamic filter. The orders of identified blocks are
unknown, and can be infinite. We recall basic facts concerning the cross-
correlation method for estimation of the impulse response of the linear dynamic objects and kernel regression or orthogonal expansion methods for
estimation of nonlinear static blocks.

1.2.1 Cross-Correlation Analysis

Let us consider the general description (without knowledge of the difference
equation and its orders) of the linear dynamic block with the infinite impulse
response $\{\lambda_i\}_{i=0}^{\infty}$, i.e.,

$$y_k = v_k + \varepsilon_k, \qquad v_k = \sum_{i=0}^{\infty} \lambda_i u_{k-i}.$$

where the excitation $\{u_k\}$ is i.i.d. random process, and for simplicity of presentation $Eu_k = 0$. We have that

$$E\{u_k y_{k+p}\} = E\left\{u_k \left(\sum_{i=0}^{\infty} \lambda_p u_{k+p-i}\right)\right\} = \beta \lambda_p,$$

where $\beta = \mathrm{var} u_k = \mathrm{const}$. Since $\lambda_p = \frac{E\{u_k y_{k+p}\}}{\mathrm{var} u_k}$, the natural estimate is

$$\widehat{\lambda}_p = \frac{\frac{1}{N}\sum_{k=1}^{N} u_k y_{k+p}}{\frac{1}{N}\sum_{k=1}^{N} u_k^2}. \tag{1.23}$$

It is shown e.g. in [128] and [50], then the convergence rate is

$$E\left(\widehat{\kappa}_p - \kappa_p\right)^2 = \mathcal{O}\left(\frac{1}{N}\right), \tag{1.24}$$

as $N \to \infty$ (asymptotically). Consequently, when the stable IIR linear sub-
system is modelled by the filter with the impulse response $\widehat{\lambda}_0, \widehat{\lambda}_1, ..., \widehat{\lambda}_{n(N)}$,
then it is free of the asymptotic approximation error if $n(N) \to \infty$ and
$n(N)/N \to 0$ as $N \to \infty$.

1.2.2 Kernel Regression Function Estimation

Now, let us consider the *SISO* static nonlinear block

$$y_k = w_k + z_k, \qquad w_k = \mu(u_k), \tag{1.25}$$

presented in Fig. 1.5, where $\mu()$ is an unknown static characteristic to be identified, and z_k is zero-mean ergodic random disturbance, independent of the input u_k. Emphasize that, if we can fix $u_k = u = \text{const}$ (active experiment), the output $y_k = \mu(u) + z_k$ is random, but the estimation of $\mu(u)$ can be made by a simple output averaging. Nevertheless, in the whole book we assume that the input process $\{u_k\}$ is random, and the role of experimenter is limited to collecting measurements. In this case, we note that the regression function (conditional expectation)

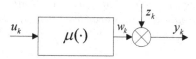

Fig. 1.5 The nonlinear *SISO* static element

$$R(u) \triangleq E\{y_k|u_k = u\}, \tag{1.26}$$

is equivalent to the identified static nonlinearity, i.e.,

$$R(u) = E\{\mu(u_k)|u_k = u\} + \underbrace{E\{z_k|u_k = u\}}_{=0} = \mu(u).$$

The estimation of $R(u) = \mu(u)$ is made with the use of N random input-output pairs $\{(u_k, y_k)\}_{k=1}^N$, assuming that the nonlinear characteristic $\mu()$ is continuous in the point u, i.e., that $|u_k - u| < h$ implies $|\mu(u_k) - \mu(u)| < H$, some $0 < H < \infty$.

In general, the kernel regression estimate has the form (see [50], [52], [155], [45])

$$\widehat{\mu}_N(u) = \widehat{R}_N(u) = \frac{\sum_{k=1}^N y_k K\left(\frac{u_k - u}{h_N}\right)}{\sum_{k=1}^N K\left(\frac{u_k - u}{h_N}\right)}, \tag{1.27}$$

where h_N is a bandwidth parameter, which fulfils the following conditions

$$h_N \to 0 \text{ and } Nh_N \to \infty, \text{ as } N \to \infty, \tag{1.28}$$

and $K()$ is a kernel function, such that

$$K(x) \geqslant 0, \ \sup K(x) < \infty \text{ and } \int K(x)dx < \infty. \tag{1.29}$$

The condition $h_N \to 0$ assures that the bias error of $\widehat{\mu}_N(u)$ tends to zero, and $N h_N \to \infty$ guarantees that more and more measurements falls into the narrowing neighborhood of u, i.e., also the variance of the estimate asymptotically tends to zero. Standard examples of the kernel function and bandwidth parameter are $K(x) = I_{[-0.5, 0.5]}(x) \triangleq \begin{cases} 1, & \text{as } |x| \leqslant 0.5 \\ 0, & \text{elsewhere} \end{cases}$, $(1 - |x|) I_{[-1,1]}(x)$ or $\left(1/\sqrt{2\pi} \right) e^{-x^2/2}$ and $h_N = h_0 N^{-\alpha}$ with $0 < \alpha < 1$ and positive $h_0 = $ const. The most important properties of the estimate (5.55) are given in the Remark below.

Remark 1.1. *[50] If the characteristic $\mu()$ and the input probability density function $f()$ are continuous in the point u, then, it holds that*

$$\widehat{\mu}_N(u) \to \mu(u) \tag{1.30}$$

in probability, as $N \to \infty$, provided that $f(u) > 0$. If moreover $\mu()$ and $f()$ are at least two times continuously differentiable at u, then for $h_N = h_0 N^{-1/5}$ the convergence rate is $|\widehat{\mu}_N(u) - \mu(u)| = O(N^{-2/5})$ in probability.

1.2.3   Orthogonal Expansion

Second kind of nonparametric regression methods are based on orthogonal series expansion. Let us define

$$g(u) \triangleq R(u) f(u) \tag{1.31}$$

and rewrite the regression $R(u)$ as

$$R(u) = \frac{g(u)}{f(u)}. \tag{1.32}$$

Let $\{\varphi_i(u)\}_{i=0}^{\infty}$ be the complete set of orthonormal functions in the input domain. If $g()$ and $f()$ are square integrable, the orthogonal series representations of $g(u)$ and $f(u)$ in the basis $\{\varphi_i(u)\}_{i=0}^{\infty}$ has the form

$$g(u) = \sum_{i=0}^{\infty} \alpha_i \varphi_i(u), \qquad f(u) = \sum_{i=0}^{\infty} \beta_i \varphi_i(u), \tag{1.33}$$

with the infinite number of coefficients $(i = 0, 1, ..., \infty)$

$$\alpha_i = E y_k \varphi_i(u_k), \qquad \beta_i = E \varphi_i(u_k). \tag{1.34}$$

Since the natural estimates of α_i's and β_i's are

$$\widehat{\alpha}_{i,N} = \frac{1}{N} \sum_{k=1}^{N} y_k \varphi_i(u_k), \qquad \widehat{\beta}_{i,N} = \frac{1}{N} \sum_{k=1}^{N} \varphi_i(u_k), \tag{1.35}$$

we obtain the following ratio estimate of $\mu(u)$

$$\widehat{\mu}_N(u) = \widehat{R}_N(u) = \frac{\sum_{i=0}^{q(N)} \widehat{\alpha}_{i,N} \varphi_i(u)}{\sum_{i=0}^{q(N)} \widehat{\beta}_{i,N} \varphi_i(u)}, \tag{1.36}$$

where $q(N)$ is some cut-off (approximation) level [40]. The consistency conditions, with respect to $q(N)$, are given in Remark below. Optimal choice of $q(N)$ with respect to the rate of convergence is considered, e.g., in [40] and [69].

Remark 1.2. *[50] To assure vanishing of the approximation error, the scale $q(N)$ must behave so that $\lim_{N \to \infty} q(N) = \infty$. For the convergence of $\widehat{\mu}_N(u)$ to $\mu(u)$, the rate of $q(N)$-increasing must be appropriately slow, e.g., $\lim_{N \to \infty} q^2(N)/N = 0$ for trigonometric or Legendre series, $\lim_{N \to \infty} q^6(N)/N = 0$ for Laguerre series, $\lim_{N \to \infty} q^{5/3}(N)/N = 0$ for Hermite series.*

1.3 Example

In Fig. 1.6 we show the results of simple experiment. The aim is to present the difference between parametric and nonparametric philosophy in system identification. The static nonlinear element $y = \text{sgn}(u) + u$ was excited by the uniformly distributed random sequence $u_k \sim U[-1, 1]$, $(k = 1, 2, ..., N)$, and its output was corrupted by the noise $z_k \sim U[-1, 1]$. The fifth order polynomial model obtained by the least squares method is compared to the kernel regression estimate with the window kernel and $h(N) = \frac{1}{\sqrt{N}}$. The figure illustrates the fact that, for N small, nonparametric kernel (local) regression estimate is very sensitive on the output noise, whereas the parametric methods can give satisfactory models, particularly if the assumed class is correct. On the other hand, if the number of measurements is large enough, nonparametric estimates are able to recover the true characteristic without any prior assumptions.

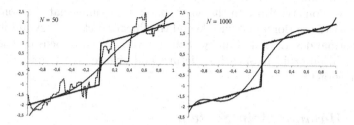

Fig. 1.6 The true characteristic (thick line), polynomial model (thin line) and the nonparametric kernel model (dashed line) for $N = 50$ and $N = 1000$

1.4 Block-Oriented Systems

Conception of the block-oriented models (interconnections of static nonlinearities and linear dynamics) in system identification has been introduced in 1980's by Billings and Fakhouri ([13]), as an alternative for Volterra series expansions (see [53], [119]). It was commonly accepted because of satisfactory approximation capabilities of various real processes and relatively small model complexity. The decentralized approach to the identification of block-oriented complex systems seems to be most natural and desirable, as such an approach corresponds directly to the own nature of systems composed of individual elements distinguished in the structure ([3], [7], [19], [101]) and tries to treat the components 'locally' as independent, autonomous objects. The Hammerstein system, built of a static non-linearity and a linear dynamics connected in a cascade, is the simplest structure in the class and hence for the most part considered in the system identification literature. Unfortunately, the popular, parametric, methods elaborated for Hammerstein system identification do not allow full decentralization of the system identification task, i.e. independent identification of a static nonlinearity and a linear dynamics – first of all, because of inaccessibility for measurements of the inner interconnection signal. They assume that the description of system components, i.e., of a static nonlinearity and a linear dynamics is known up to the parameters (a polynomial model along with a FIR dynamics representation are usually used) and these parameters are "glued" when using standard input-output data of the overall system for identification purposes (e.g. [13], [160], [131], [130]). On the other hand, in a nonparametric setting (the second class of existing identification methods, see, e.g. [48], [49], [107]) no preliminary assumptions concerning the structure of subsystems are used and only the data decide about the obtained characteristics of the system components but then any possible a priori knowledge about the true description of subsystems is not exploited, i.e. inevitably lost.

Below, we present the most popular structures of block-oriented systems [36], i.e., Hammerstein system, Wiener system, Wiener-Hammerstein (sandwich) system, additive NARMAX system, and finally, the system with arbitrary connection structure. In the whole book we assume that the nonlinear characteristics are Borel measurable and the linear blocks are asymptotically stable. Formal description of all systems is given. We also focus on specifics of identification problem of each structure, and give some examples of applications in practice.

1.4.1 Hammerstein System

The Hammerstein system, shown in Fig. 1.7, consists of a static nonlinear block with the unknown characteristic $\mu()$, followed by the linear dynamic object with the unknown impulse response $\{\gamma_i\}_{i=0}^{\infty}$. The internal signal

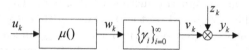

Fig. 1.7 Hammerstein system

$w_k = \mu(u_k)$ is not accessible, and the output is observed in the presence of the random noise z_k.

$$y_k = \sum_{i=0}^{\infty} \gamma_i \mu(u_{k-i}) + z_k \qquad (1.37)$$

The aim is to estimate both $\mu()$ and $\{\gamma_i\}_{i=0}^{\infty}$ on the basis of N input output measurements $\{(u_k, y_k)\}_{k=1}^{N}$ of the whole Hammerstein system. Obviously, since the signal w_k is hidden, both elements can de identified only up to some scale factor (the system with the characteristic $c\mu()$ and linear dynamics $\{\frac{1}{c}\gamma_i\}_{i=0}^{\infty}$ is not distinguishable with the original one, from the input-output point of view). This problem cannot be avoided without the additional knowledge of the system, and is independent of the identification method. In the literature it is often assumed, without any loss of generality, that $\gamma_0 = 1$.

First attempts to Hammerstein system identification ([101], [19], [12]-[13]), proposed in 1970s and 1980s, assumed polynomial form of $\mu()$ and FIR linear dynamics with known orders p and m. The system was usually parametrized as follows

$$w_k = c_p u_k^p + c_{p-1} u_k^{p-1} + ... + c_1 u + c_0, \qquad (1.38)$$

$$y_k = \sum_{i=0}^{m} b_i w_{k-i} + z_k. \qquad (1.39)$$

Since the nonlinear characteristic (1.38) is linear with respect to the parameters, and the second subsystem is linear, also the whole Hammerstein system is linear with respect to parameters and can be described in the following form

$$y_k = \phi_k \theta + z_k, \qquad (1.40)$$

where the regressor ϕ_k includes all combinations of u_{k-i}^l ($i = 0, 1, ..., m$, and $l = 0, 1, ..., p$), and the vector θ is built of all respective mixed products of parameters, i.e., $c_l b_i$ ($i = 0, 1, ..., m$, and $l = 0, 1, ..., p$). Such a description allows for application of the linear least squares method for estimation of θ in (1.40). Nevertheless, using ordinary polynomials leads to the tasks, which are very badly conditioned numerically. Moreover, if the assumed parametric model is bad, the approximation error appears, which cannot be reduced by increasing number of data [121].

As regards the nonparametric methods ([100], [40]-[49]), strongly elaborated in 1980s, they are based on the observation that, for i.i.d. excitation $\{u_k\}$, the input-output regression in Hammerstein system is equivalent to the nonlinear characteristic of the static element, up to some constant δ, i.e.,

$$R(u) = E\{y_k|u_k = u\} = E\left\{\gamma_0\mu(u_k) + \sum_{i=1}^{\infty}\gamma_i\mu(u_{k-i})|u_k = u\right\} =$$

$$= \gamma_0\mu(u_k) + \delta,$$

where the offset $\delta = \sum_{i=1}^{\infty}\gamma_i E\mu(u_1)$ can be simply computed if the characteristic is known in at least one point, e.g., we know that $\mu(0) = 0$. As it was shown in e.g. [48] and [50], standard nonparametric regression estimates $\hat{R}_N(u)$ (kernel or orthogonal) can be successfully applied for correlated data $\{(u_k, y_k)\}_{k=1}^{N}$. Under some general conditions, they recover the shape of true nonlinear characteristic, and are asymptotically free of approximation error. The cost paid for neglecting the prior knowledge is large variance of the model for small number of measurements. The reason of the variance is that the Hammerstein system is in fact treated as the static element, and the 'historical' term $\sum_{i=1}^{\infty}\gamma_i\mu(u_{k-i})$ is treated as the 'system' noise. Hence, nonparametric methods are better choice as regards the asymptotic properties, when the number of measurements is large enough (required number obviously depends on specifics of the system).

1.4.2   Wiener System

Since in the Wiener system the nonlinear block precedes the linear dynamics, the identification task is much more difficult. Till now, sufficient identifiability conditions have been formulated only fore some special cases. Many methods require the non-linearity to be known, invertible, differentiable or require special input sequences (see e.g. [12], [61], [54]). The discrete-time Wiener system (see Fig. 1.8), i.e. the linear dynamics with the impulse response $\{\lambda_j\}_{j=0}^{\infty}$, connected in a cascade with the static nonlinear block, characterized by $\mu()$, is described by the following equation

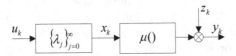

Fig. 1.8 Wiener system

$$y_k = \mu \left(\sum_{j=0}^{\infty} \lambda_j u_{k-j} \right) + z_k, \qquad (1.41)$$

where u_k, y_k, and z_k are the input, output and the random disturbance, respectively. The goal of identification is to recover characteristics of both elements, i.e. $\left\{ \widehat{\lambda}_j \right\}_{j=0}^{\infty}$ and $\widehat{\mu}(x)$ for each $x \in R$, using the set of input-output measurements $\{(u_k, y_k)\}_{k=1}^{N}$. In the traditional (parametric) approach we also assume finite dimensional functional, e.g. the ARMA-type dynamic block

$$x_k + a_1^* x_{k-1} + ... + a_r^* x_{k-r} = b_0^* u_k + b_1^* u_{k-1} + ... + b_s^* u_{k-s},$$
$$x_k = \phi_k^T \theta^*, \qquad (1.42)$$
$$\phi_k = (-x_{k-1}, ..., -x_{k-r}, u_k, u_{k-1}, ..., u_{k-s})^T,$$
$$\theta^* = (a_1^*, ..., a_r^*, b_0^*, b_1^*, ..., b_s^*)^T,$$

and given formula $\mu(x, c^*) = \mu(x)$ including finite number of unknown true parameters $c^* = (c_1^*, c_2^*, ..., c_m^*)^T$. Respective Wiener model is thus represented by $r + (s + 1) + m$ parameters, i.e.,

$$\overline{x}_k \triangleq \overline{\phi}_k^T \theta, \text{ and } \overline{x}_k = 0 \text{ for } k \leqslant 0,$$
$$\text{where } \overline{\phi}_k^T = (-\overline{x}_{k-1}, ..., -\overline{x}_{k-r}, u_k, u_{k-1}, ..., u_{k-s})^T, \qquad (1.43)$$
$$\theta = (a_1, ..., a_r, b_0, b_1, ..., b_s)^T,$$
$$\text{and } \overline{y}(x, c) = \mu(x, c), \text{ where } c = (c_1, c_2, ..., c_m)^T.$$

If x_k had been accessible for measurements then the true system parameters could have been estimated by the following minimizations

$$\widehat{\theta} = \arg\min_{\theta} \sum_{k=1}^{N} (x_k - \overline{x}_k(\theta))^2, \qquad \widehat{c} = \arg\min_{c} \sum_{k=1}^{N} (y_k - \overline{y}(x_k, c))^2. \quad (1.44)$$

Here we assume that only the input-output measurements (u_k, y_k) of the whole Wiener system are accessible, and the internal signal x_k is hidden. This feature leads to the following nonlinear least squares problem

$$\widehat{\theta}, \widehat{c} = \arg\min_{\theta, c} \sum_{k=1}^{N} [y_k - \overline{y}(\overline{x}_k(\theta), c)]^2, \qquad (1.45)$$

which is usually very complicated. Moreover, uniqueness of the solution in (1.45) cannot be guaranteed in general, as it depends on both input distribution, types of models, and values of parameters.

For example, the Wiener system with the polynomial static characteristic and the *FIR* linear dynamics

$$y_k = \sum_{i=0}^{p} c_0 w_k^i + z_k, \qquad w_k = \sum_{i=0}^{m} \gamma_i u_{k-i}, \qquad (1.46)$$

can be described as

$$Y_N = \Phi_N \theta + Z_N, \qquad (1.47)$$

but now, the meaning and the structure of the matrix $\Phi_N = (\phi_1, ..., \phi_N)^T$ and the vector θ are more sophisticated, i.e.,

$$\phi_k = \left[\frac{|\alpha|!}{\alpha!} \overline{u}_k^\alpha \right]_{|\alpha| \leqslant p} \in \mathcal{R}^d, \qquad (1.48)$$

$$\theta = \left[c_{|\alpha|} \Gamma^\alpha \right]_{|\alpha| \leqslant p} \in \mathcal{R}^d,$$

where

$$\alpha = (\alpha_1, \alpha_2, ..., \alpha_{m+1})^T \in \mathcal{N}_0^{m+1},$$

is the multi-index of order $m + 1$ (see [81]), $|\alpha| = \sum_{i=0}^{m+1} \alpha_i$, $\alpha! = \prod_{i=1}^{m+1} \alpha_i$, $\overline{u}_k = (u_k, u_{k-1}, ..., u_{k-m})^T$, $\overline{u}_k^\alpha = \prod_{i=1}^{m+1} u_{k-i-1}^{\alpha_i}$, $\Gamma^\alpha = \prod_{i=1}^{m+1} \gamma_{i-1}^{\alpha_i}$, $d = \sum_{i=0}^{p} \frac{(m+i)!}{m! i!}$, and $[f(\alpha)]_{|\alpha| \leqslant p}$ denotes the column vector whose components are evaluated at every multi-index α such that $|\alpha| \leqslant p$ under some established ordering. Estimate of θ can be computed by the standard least squares, analogously as for Hammerstein system, but extraction of parameters of components requires application of multi-dimensional singular value decomposition (SVD) procedure.

As regards nonparametric methods, proposed in 1980s, for Wiener system identification, they were usually based on the restrictive assumptions that the nonlinear characteristic is invertible or locally invertible and the input process is Gaussian. Moreover, existence of output random noise were not admitted. It was shown, that under above conditions the reverse regression function

$$R^{-1}(y) = E\left(u_k | y_{k+p} = y\right) = \alpha_p \mu^{-1}(y)$$

is equivalent to the inversion of the identified characteristic up to some (impossible to identify) scale α_p. Recently, several new ideas was proposed in the literature (see e.g. [44], [90], [108]), admitting arbitrary input density, nonivertible characteristic and IIR linear block. Nevertheless, the rate of convergence of estimates is still not satisfying.

1.4.3 Sandwich (Wiener-Hammerstein) System

In the Wiener-Hammerstein (sandwich) system, presented in Fig. 1.9, the nonlinear block $\mu()$ is surrounded by two linear dynamics $\{\lambda_j\}_{j=0}^{\infty}$ and $\{\gamma_i\}_{i=0}^{\infty}$. It can be described as follows

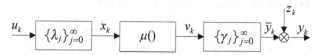

Fig. 1.9 Wiener-Hammerstein (sandwich) system

$$y_k = \sum_{i=0}^{\infty} \gamma_i v_{k-i} + z_k, \text{ where } v_k = \mu \left(\sum_{j=0}^{\infty} \lambda_j u_{k-j} \right). \tag{1.49}$$

Since both x_k and v_k cannot be measured, the system as a whole cannot be distinguished with the system composed with the elements $\{\frac{1}{c_1}\lambda_j\}$, $c_2\mu(c_1x)$, $\{\frac{1}{c_2}\gamma_i\}$, and the nonlinear characteristic can be identified only up to c_1 and c_2. Analogously as for Hammerstein and Wiener systems, without any loss of generality, one can assume that $\gamma_0 = 1$ and $\lambda_0 = 1$. The Hammerstein system and Wiener systems are special cases of (1.49), obtained for $\lambda_j = 1, 0, 0, ...$ or $\gamma_i = 1, 0, 0, ...$, respectively.

1.4.4 NARMAX System

We also consider in this book a special case of NARMAX systems, consisting of two branches of Hammerstein structures, from which one plays the role of nonlinear feedback (see Fig. 1.10). The system is described by the equation

$$y_k = \sum_{j=1}^{\infty} \lambda_j \eta(y_{k-j}) + \sum_{i=0}^{\infty} \gamma_i \mu(u_{k-i}) + z_k, \tag{1.50}$$

where $\eta()$ is nonlinear static characteristic in feedback. The system is assumed to be stable as a whole. As it was shown in the Appendix A.3, the Hammerstein system is a special case of (1.50) when the characteristic $\eta()$ is linear.

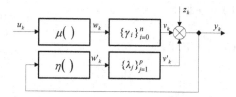

Fig. 1.10 The additive NARMAX system

1.4.5 Interconnected MIMO System

Finally, we introduce the system with arbitrary structure of connections (Fig. 1.11). It consists of n blocks described by unknown functional $F_i(\{u_i\}, \{x_i\})$, $i = 1, 2, ..., n$. Only external inputs u_i and outputs y_i of the system can be measured. The interactions x_i are hidden, but the structure of connections is known and coded in the zero-one matrix H, i.e.,

$$x_i = H_i \cdot (y_1, y_2, ..., y_n)^T + \delta_i,$$

where H_i denotes ith row of H, and δ_i is a random disturbance. In the simplest case of static linear system, the single block can be described as follows

$$y_i = (x_i, u_i)(a_i, b_i)^T + \xi_i \qquad (i = 1, 2, ..., n),$$

where a_i and b_i are unknown parameters and ξ_i is a random output noise. In more general case of nonlinear and dynamic system, the single block can be represented (approximated) by, e.g., two channels of Hammerstein models (see Fig. 1.12), resembling the Narmax/Lur'e system, considered in Chapter 5.

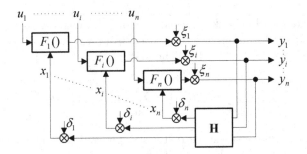

Fig. 1.11 The system with arbitrary structure

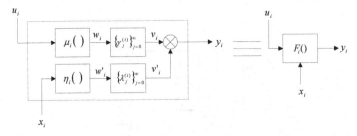

Fig. 1.12 Nonlinear dynamic model of the single block in complex system

1.4.6 Applications in Science and Technology

Block-oriented models are often met in various domains of science and technology. For details, we refer the reader to the paper [35], including a huge number of publications on this topic. Examples of the most popular applications are given below:

- signal processing (time series modeling and forecasting, noise reduction, data compression, EEG analysis),
- automation and robotics (fault detection, adaptive control),
- telecommunication and electronics (channel equalization, image compression, diode modeling, testing of AM and FM decoders),
- acoustics (noise and echo cancellation, loudspeaker linearization),
- biocybernetics (artificial eye, artificial muscle, model of neuron),
- geology (anti-flood systems, water level modeling, deconvolution),
- chemistry (modeling of distillation and fermentation processes, pH neutralization),
- physics (adaptive optics, heat exchange processes, modeling of the Diesel motor).

Chapter 2
Hammerstein System

2.1 Finite Memory Hammerstein System

In this Section, a parametric-nonparametric, methodology for Hammerstein system identification, introduced in [64] and [67], is reminded. Assuming white random input and correlated output noise, the parameters of a nonlinear static characteristic and FIR system dynamics are estimated separately, each in two stages. First, the inner signal is recovered by a nonparametric regression estimation method (Stage 1) and then, system parameters are solved independently by the least squares (Stage 2). Convergence properties of the scheme are shown and rates of convergence are given.

2.1.1 Introduction

As it was mentioned before, traditional methods assume prior knowledge of the system components up to a finite number of parameters (e.g. a polynomial form of the nonlinear static characteristic $\mu(u)$; see, for example [6], [38],[54] and the references therein) while the latter do not impose any specific structure on the system description (e.g. [40], [48], [70]). In this Chapter parametric-nonparametric technique is applied to cope with the Hammerstein system identification task. Like in a parametric setting, we assume that the nonlinear static characteristic $\mu(u)$ is known with accuracy to the parameters and, further, that the linear dynamic block is a FIR filter of the known order. In the proposed approach the identification is performed in two stages. First, exploiting a nonparametric regression estimation technique, the unmeasurable inner signal $\{w_k\}$ is estimated from the measurement data (u_k, y_k). Then, the least squares method is used to the independent estimation of the two subsystems parameters using, respectively, the pairs $(u_k, \widehat{w}_k)$ and $(\widehat{w}_k, y_k)$ where $\{\widehat{w}_k\}$ is the estimate of the interaction sequence obtained by a nonparametric method. As compared to the parametric identification techniques developed to date, the potential advantages of the approach are that:

G. Mzyk, *Combined Parametric-Nonparametric Identification*
of Block-Oriented Systems, Lecture Notes in Control and Information Sciences 454,
DOI: 10.1007/978-3-319-03596-3_2, © Springer International Publishing Switzerland 2014

- we get simple estimates of both subsystems, given by the explicit formulas,
- the routine is not using any type of alternate updating,
- the method works with systems having nonpolynomial static characteristics,
- the algorithm operates efficiently for both white and colored noise, without the need of recovering the noise model,
- each part of the system is identified separately making the estimates robust against lack or falsity of a priori information about the other part, and
- convergence properties are proved and rates of convergence are given.

We refer to [101], [56], [13], [2], [144] and the references cited therein for representative examples of the parametric identification methods worked out for the Hammerstein system and to [54] for a comprehensive overview of the subject.

2.1.2 Statement of the Problem

In the Hammerstein system in Fig. 2.1, u_k, y_k and z_k are respectively the input, output, and noise processes at instant k, and w_k is the internal signal (interaction input) not accessible for measurement (see [54] for a discussion). We assume the following

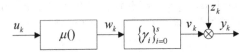

Fig. 2.1 Finite memory Hammerstein system

Assumption 2.1. *The nonlinear static characteristic $\mu(u)$ is known up to a finite number of parameters $c_1, ..., c_m$ and is a generalized polynomial [23]*

$$\mu(u) = \sum_{l=1}^{m} c_l f_l(u) \tag{2.1}$$

where $f_1(u), ..., f_m(u)$ is a set of known linearly independent basis functions, such that

$$|f_l(u)| \leqslant p_{max}; \quad l = 1, 2, ..., m \tag{2.2}$$

some $p_{max} > 0$ for u in the operation region $|u| \leq u_{max}$.

Assumption 2.2. *Linear dynamics is a FIR filter,*

$$v_k = \sum_{i=0}^{s} \gamma_i w_{k-i} \tag{2.3}$$

some finite known order s, with the unknown impulse response $\{\gamma_i\}_{i=0}^{s}$.

Assumption 2.3. *The input signal $\{u_k\}$ is a bounded i.i.d. random process, $|u_k| \leq u_{max}$, some $u_{max} > 0$.*

Assumption 2.4. *The output noise $\{z_k\}$ is a correlated process generated from a bounded zero-mean white noise $\{\varepsilon_k\}$ ($E\varepsilon_k = 0$, $|\varepsilon_k| \leq \varepsilon_{max}$, some $\varepsilon_{max} > 0$) by an asymptotically stable linear filter with unknown impulse response $\{\omega_i\}_{i=0}^{\infty}$ ($\sum_{i=0}^{\infty} |\omega_i| < \infty$),*

$$z_k = \sum_{i=0}^{\infty} \omega_i \varepsilon_{k-i}, \qquad (2.4)$$

independent of the input signal $\{u_k\}$. Consequently, $\{z_k\}$ is a zero-mean and bounded process; $Ez_k = 0$, $|z_k| \leq z_{max}$, where $z_{max} = \varepsilon_{max} \sum_{i=0}^{\infty} |\omega_i| < \infty$.

Assumption 2.5. *$\mu(u_0)$ is known at some point u_0 and $\gamma_0 = 1$.*

The last requirement is strictly connected with the presented method and will be explained in Section 2.1.3.

Owing to (2.1), (2.3) and the relationship $w_k = \mu(u_k)$ we get

$$w_k = \phi^T(u_k) c, \quad y_k = \vartheta_k^T \gamma + z_k \qquad (2.5)$$

where

$$c = (c_1, c_2, ..., c_m)^T, \qquad \gamma = (\gamma_0, \gamma_1, ..., \gamma_s)^T,$$
$$\phi(u_k) = (f_1(u_k), f_2(u_k), ..., f_m(u_k))^T, \quad \text{and} \quad \vartheta_k = (w_k, w_{k-1}, ..., w_{k-s})^T.$$

Our objective is to recover unknown c and γ, using input-output measurements $\{(u_k, y_k)\}_{k=1}^{M}$ of the *whole* Hammerstein system.

2.1.3 Background of the Approach

Denoting

$$\Phi_{N_0} = (\phi(u_1), \phi(u_2), ..., \phi(u_{N_0}))^T,$$
$$W_{N_0} = (w_1, w_2, ..., w_{N_0})^T, \quad \Theta_N = (\vartheta_{1+s}, \vartheta_{2+s}, ..., \vartheta_{N+s})^T,$$
$$Y_N = (y_{1+s}, y_{2+s}, ..., y_{N+s})^T, \quad Z_N = (z_{1+s}, z_{2+s}, ..., z_{N+s})^T,$$

from (2.5) we have

$$W_{N_0} = \Phi_{N_0} c, \quad Y_N = \Theta_N \gamma + Z_N. \qquad (2.6)$$

Hence, we can get the following least squares solution for the parameter vector c,

$$c = \left(\Phi_{N_0}^T \Phi_{N_0}\right)^{-1} \Phi_{N_0}^T W_{N_0}, \qquad (2.7)$$

and the least squares estimate of the vector γ,

$$\widehat{\gamma}_N = (\Theta_N^T \Theta_N)^{-1} \Theta_N^T Y_N \tag{2.8}$$

(weakly consistent in the problem in question [128]), provided that Φ_{N_0} and Θ_N are of full column rank. To this end, let us assume that $N_0 \geq m = \dim c$ and $N \geq s + 1 = \dim \gamma$. The obvious difficulty in carrying out this simple approach is that the w_k's entering W_{N_0} and Θ_N cannot be measured. A way to overcome this drawback and utilize the solutions (2.7) and (2.8) is to recover the unknown w_k's from the measurement data $\{(u_k, y_k)\}_{k=1}^M$ and to use the obtained estimates $\widehat{w}_{k,M}$ instead of w_k. Observing that (see (2.3), (2.4) along with the relations $w_k = \mu(u_k)$ and $y_k = v_k + z_k$)

$$y_k = \gamma_0 \mu(u_k) + \sum_{i=1}^{s} \gamma_i \mu\left(u_{k-i}\right) + \sum_{i=0}^{\infty} w_i \varepsilon_{k-i},$$

or equivalently

$$y_k = (\gamma_0 \mu(u_k) + d) + \sum_{i=1}^{s} \gamma_i \left[\mu\left(u_{k-i}\right) - E\mu\left(u_1\right)\right] + \sum_{i=0}^{\infty} w_i \varepsilon_{k-i}, \tag{2.9}$$

where $d = E\mu\left(u_1\right) \sum_{i=1}^{s} \gamma_i$ and viewing the tail in equation (2.9) as the aggregate zero-mean output noise, we however realize that only scaled and shifted version $\mu_b(u) = \gamma_0 \mu(u) + d$ of the nonlinear characteristic $\mu(u)$ is at most accessible from the input-output data $\{(u_k, y_k)\}$, and hence only scaled and shifted values $\gamma_0 w_k + d$ of the interactions $w_k = \mu(u_k)$ can be recovered in a general case. To get consistent estimates $\widehat{w}_{k,M}$ of w_k, we require the technical Assumption 2.5. Then, for $\gamma_0 = 1$, the bias d can be clearly eliminated by taking, for a given u_k, the estimate $\widehat{w}_{k,M} = \widehat{\mu}_{b,M}(u_k) - \widehat{\mu}_{b,M}(u_0) + \mu_0$ where $\widehat{\mu}_{b,M}(u)$ is an estimate of the characteristic $\mu_b(u)$. While the knowledge of $\mu(u)$ in at least one point u_0, $\mu(u_0) = \mu_0$, is necessary for removing the bias (unless $E\mu\left(u_1\right) = 0$ or $\sum_{i=1}^{s} \gamma_i = 0$) and this requirement cannot be dropped or relaxed, the requirement that $\gamma_0 = 1$ can be weakened to the demand that $\gamma_0 \neq 0$ as the parameter sets $(\gamma_0 c, (1/\gamma_0) \gamma)$ and (c, γ) are equivalent (indistinguishable) from the input-output data point of view for each $\gamma_0 \neq 0$ (see (2.9) and [4], [5] for the discussion of this issue). Introducing in particular the regression function

$$R(u) = E\{y_k \mid u_k = u\} \tag{2.10}$$

we get for $\gamma_0 = 1$ that $R(u) = \mu(u) + d$ (cf. (2.9) and Assumptions 2.1 and 2.4) and further $R(u) - R(u_0) = \mu(u) - \mu(u_0)$ or $\mu(u) = R(u) - R(u_0) + \mu(u_0)$. For technical reasons, without any loss of generality, we assume henceforth $u_0 = 0$ and $\mu(0) = 0$ getting

$$\mu(u) = R(u) - R(0) \tag{2.11}$$

The fundamental formula (2.11) relating the nonlinearity $\mu(u)$ with the regression function $R(u)$ was given in [48]. We note that $R(u) = \mu_b(u)$.

Equations (2.10) and (2.11) along with $w_k = \mu(u_k)$ suggest that estimation of interactions w_k from the data $\{(u_k, y_k)\}_{k=1}^{M}$ can in particular be performed by a nonparametric regression function estimation method, without the use of parametric prior knowledge of the system components. The class of nonparametric regression function estimates elaborated for the Hammerstein system to date comprises kernel estimates [48], [49], orthogonal series estimates [40], and wavelet estimates [69], [70], [125]. For a general treatment of nonparametric regression function estimation methods see, for example, [30], [50], [59], [60], [149].

Remark 2.1. *The idea of recovering unmeasurable inner signal in the Hammerstein and Wiener system has been realized in [4], [5] by using frequency domain method and a nonparametric method to support the recovery from independent data of the linear regression parameters was used in [25].*

2.1.4 Two-Stage Parametric-Nonparametric Estimation of Nonlinearity Parameters

Due to the above, we propose the following two-stage estimation procedure of vector c.

Stage 1 (nonparametric): Using input-output data $\{(u_k, y_k)\}_{k=1}^{M}$, for a set of input points $\{u_n; \; n = 1, 2, ..., N_0\}$ such that $M > N_0 \geq m = \dim c$ and $\Phi_{N_0} = (\phi(u_1), \phi(u_2), ..., \phi(u_{N_0}))^T$ is of full column rank, estimate the corresponding interactions $\{w_n = \mu(u_n); \; n = 1, 2, ..., N_0\}$ as

$$\widehat{w}_{n,M} = \widehat{R}_M(u_n) - \widehat{R}_M(0), \tag{2.12}$$

where $\widehat{R}_M(u)$ is a nonparametric estimate of the regression function $R(u)$, computed for $u \in \{0, u_n; \; n = 1, 2, ..., N_0\}$. Here, the estimation points $\{u_n\}_{n=1}^{N_0}$ can be the measured input data or the points freely selected by the experimenter; we do not distinguish these two situations because of no formal importance of the difference.

Stage 2 (parametric): Compute the estimate of parameter vector c as (cf. (2.7))

$$\widehat{c}_{N_0,M} = \left(\Phi_{N_0}^T \Phi_{N_0}\right)^{-1} \Phi_{N_0}^T \widehat{W}_{N_0,M} \tag{2.13}$$

where Φ_{N_0} and $\widehat{W}_{N_0,M} = (\widehat{w}_{1,M}, \widehat{w}_{2,M}, ..., \widehat{w}_{N_0,M})^T$ are established in Stage 1.

Remark 2.2. *The requirement that $\mathrm{rank}\Phi_{N_0} = m = \dim c$ can be fulfilled because of linear independence of $f_1(u), ..., f_m(u)$ and the fact that the estimation points u_n in Stage 1 may in particular be selected in an arbitrary manner. Such a condition is for instance automatically satisfied if $N_0 \geq m$,*

$u_1, u_2, ..., u_{N_0}$ are distinct points, and $f_1(u), ..., f_m(u)$ in (2.1) is a Tchebycheff system (i.e., satisfies the Haar condition). Appropriate examples are $\{1, u, u^2, ..., u^{m-1}\}$ (which yields standard polynomial characteristic in (2.1)), $\{1, \sin u, \sin 2u, ..., \sin(m-1)u\}$ (on $[0, 2\pi]$) or $\{e^{\lambda_1 u}, e^{\lambda_2 u}, ..., e^{\lambda_m u}\}$ and non-degenerate linear combinations of these functions [23].

Defining the estimation error in Stage 1 as $\varsigma_{n,M} = \widehat{w}_{n,M} - w_n$ and including (2.5), we see that Stage 2 may be considered as the identification of a static element $\widehat{w}_{n,M} = \phi^T(u_n) c + \varsigma_{n,M}$ from the 'data' $\{(u_n, \widehat{w}_{n,M})\}_{n=1}^{N_0}$ by means of least squares. We note that estimation of vector c in (2.13) is performed without the use of a priori knowledge of the system dynamics, in contrast to the alternate updating methods [56], [101], [144].

The following theorem holds.

Theorem 2.1. *If for $u \in \{0, u_n; \ n = 1, 2, ..., N_0\}$ it holds that $\widehat{R}_M(u) \to R(u)$ in probability as $M \to \infty$ (Stage 1), then $\widehat{c}_{N_0,M} \to c$ in probability as $M \to \infty$ (Stage 2).*

Proof. For the proof see Appendix A.4.   ∎

Remark 2.3. *Any kind of probabilistic convergence can be considered in Theorem 2.1. We examine convergence in probability as this particular type of convergence is usually studied in the convergence analysis of nonparametric regression function estimates for Hammerstein systems (see [40], [48], [49], [70]).*

In the next theorem establishing asymptotic rate of convergence of the estimate $\widehat{c}_{N_0,M}$, for a sequence of random variables $\{\varsigma_M\}$ and positive number sequence $\{a_M\}$ convergent to zero, $\varsigma_M = O(a_M)$ in probability means that $d_M \varsigma_M / a_M \to 0$ in probability as $M \to \infty$ for any number sequence $\{d_M\}$ tending to zero ([49], p. 140).

Theorem 2.2. *If in Stage 1 $\left| \widehat{R}_M(u) - R(u) \right| = O(M^{-\tau})$ in probability as $M \to \infty$ for each $u \in \{0, u_n; \ n = 1, 2, ..., N_0\}$, then also $\|\widehat{c}_{N_0,M} - c\| = O(M^{-\tau})$ in probability as $M \to \infty$.*

Proof. For the proof see Appendix A.5.   ∎

As we see, in the method the rate of convergence for estimating vector c is determined by the nonparametric rates for the regression function. Since, as is well known, for each nonparametric method $0 < \tau < 1/2$ (e.g. [59]), the rate $O(M^{-\tau})$ is in general of slower order than the best possible parametric rate of convergence $O(M^{-1/2})$ in probability. However, for polynomial and other smooth nonlinearities the convergence rate can be made arbitrarily close to $O(M^{-1/2})$ by applying nonparametric estimates being able to adapt to smooth functions, e.g. wavelet estimates [69], [70].

2.1.5 Two-Stage Parametric-Nonparametric Identification of Linear Dynamics

Considering (2.8), we propose similar two-stage scheme to estimate γ.

Stage 1 (nonparametric): Using input-output measurement data

$$\{(u_k, y_k)\}_{k=1}^M,$$

estimate the entries $\{w_{t-r}; t = n + s; n = 1, 2, ..., N; r = 0, 1, ..., s\}$ of Θ_N by a nonparametric method, i.e. as

$$\hat{w}_{t-r,M} = \hat{R}_M(u_{t-r}) - \hat{R}_M(0),$$

where $\{u_{t-r}\}$ are the input data points corresponding to the output measurements $\{y_t\}_{t=1+s}^{N+s}$ collected in vector Y_N (see (2.6)) and $\hat{R}_M(u)$ is a nonparametric estimate of the regression function $R(u)$ computed for $u \in \{0, u_{t-r}; t = n + s; n = 1, 2, ..., N; r = 0, 1, ..., s\}$ (cf. (2.10), (2.11), (2.12)).

Stage 2 (parametric): Compute (see (2.8))

$$\hat{\gamma}_{N,M} = (\hat{\Theta}_{N,M}^T \hat{\Theta}_{N,M})^{-1} \hat{\Theta}_{N,M}^T Y_N, \qquad (2.14)$$

where

$$\hat{\Theta}_{N,M} = (\hat{\vartheta}_{1+s,M}, \hat{\vartheta}_{2+s,M}, ..., \hat{\vartheta}_{N+s,M})^T,$$
$$\hat{\vartheta}_{t,M} = (\hat{w}_{t,M}, \hat{w}_{t-1,M}, ..., \hat{w}_{t-s,M})^T,$$

and Y_N is a noisy output vector.

In Stage 1 we assume that $M - s > N \geq s + 1 = \dim \gamma$. We accentuate that estimation of γ in (2.14) is independent of estimating parameter vector c of the static subsystem.

Remark 2.4. *Recalling that the 'input' data are contaminated by the additive (estimation) errors, $\hat{w}_{t-r,M} = w_{t-r} + \varsigma_{t-r,M}$, we note that recovering of γ from the data $(\hat{\Theta}_{N,M}, Y_N)$ is in fact the errors-in-variables estimation problem. This is unlike the aforementioned [25] where input data are accurate and only output measurements are corrupted by the noise (white).*

The following theorem holds.

Theorem 2.3. *If for $u \in \{0, u_{t-r}; t = n + s; n = 1, 2, ..., N; r = 0, 1, ..., s\}$ the estimate $\hat{R}_M(u)$ is bounded and the asymptotic nonparametric estimation error in Stage 1 behaves like*

$$\left| \hat{R}_M(u) - R(u) \right| = O(M^{-\tau}) \text{ in probability}, \qquad (2.15)$$

then $\hat{\gamma}_{N,M} \to \gamma$ in probability in Stage 2 provided that $N, M \to \infty$ and $NM^{-\tau} \to 0$.

Proof. For the proof see Appendix A.6.   ■

The condition $NM^{-\tau} \to 0$ is fulfilled if $M = const \cdot N^{(1+\alpha)/\tau}$, or equivalently $N = const \cdot M^{\tau/(1+\alpha)}$, any $\alpha > 0$. It is noteworthy that since in general $0 < \tau < 1/2$, to get consistency of the estimate $\widehat{\gamma}_{N,M}$ far more data points $\{(u_k, y_k)\}_{k=1}^{M}$ must be used in Stage 1 for nonparametric estimation of interactions $\{w_{t-r}\}$ than the "observations" $\{(\widehat{\vartheta}_{t,M}, y_t)\}_{t=1+s}^{N+s}$ in Stage 2 for computing $\widehat{\gamma}_{N,M}$. This can be explained by the necessity of effective reduction of the 'input' errors in $\widehat{\vartheta}_{t,M}$'s and slower convergence of nonparametric methods. From the data length M viewpoint small α are clearly preferred. In contrast, $\alpha \geq 1/2$ are desirable from the $\widehat{\gamma}_{N,M}$ convergence rate point of view.

Theorem 2.4. *For $M \sim N^{(1+\alpha)/\tau}$, equivalently $N \sim M^{\tau/(1+\alpha)}$, $\alpha > 0$, the asymptotic convergence rate in Stage 2 is $\|\widehat{\gamma}_{N,M} - \gamma\| = O(N^{-\min(1/2,\alpha)})$ in probability.*

Proof. For the proof see Appendix A.7.   ■

If $\alpha \geq 1/2$ we attain for the estimate $\widehat{\gamma}_{N,M}$ the best possible parametric rate of convergence $O(N^{-1/2})$ in probability (w.r.t. N). This means that for $\alpha \geq 1/2$ the influence of the input (estimation) errors in $\widehat{\vartheta}_{t,M}$'s on the accuracy of the estimate $\widehat{\gamma}_{N,M}$ is dominated by the standard effect of the output measurement noise. In the case of $\alpha < 1/2$ we get slower guaranteed convergence rate of order $O(N^{-\alpha})$.

2.1.6 Example

Let us use, for example, in Stage 1 of the schemes the kernel regression estimate studied in [48], [49], [50], i.e.,

$$\widehat{R}_M(u) = \frac{\sum_{k=1}^{M} y_k K\left(\frac{u-u_k}{h(M)}\right)}{\sum_{k=1}^{M} K\left(\frac{u-u_k}{h(M)}\right)} \tag{2.16}$$

where $K(u)$ is a kernel function and $h(M)$ is a bandwidth parameter. Standard examples are $K(u) = I_{[-0.5,0.5]}(u)$, $(1-|u|)I_{[-1,1]}(u)$ or $(1/\sqrt{2\pi})\,e^{-u^2/2}$ and $h(M) = const \cdot M^{-\alpha}$ with a positive constant and $0 < \alpha < 1$; see [149]. Owing to the convergence results presented there, we find out that for each of the above-mentioned $K(u)$ it holds that $\widehat{R}_M(u) \to R(u)$ in probability as $M \to \infty$, and that the convergence takes place at every $u \in Cont(\mu, \nu)$, the set of continuity points of $\mu(u)$ and $\nu(u)$, at which $\nu(u) > 0$ where $\nu(u)$ is a probability density function of the system input (assumed to exist). Taking in particular the Gaussian kernel $K(u) = (1/\sqrt{2\pi})\,e^{-u^2/2}$ and, according to the recommendation in [49], $h(M) \sim M^{-1/5}$ we get the convergence rate $\left|\widehat{R}_M(u) - R(u)\right| = O(M^{-2/5})$ in probability and hence

$\|\widehat{c}_{N_0,M} - c\| = O(M^{-2/5})$ in probability ($\tau = 2/5$) in Stage 2 for a static element, provided that $\mu(u)$ and $\nu(u)$ are at $u \in \{0, u_n; n = 1, 2, ..., N_0\}$ at least two times continuously differentiable functions and $\nu(u) > 0$ there (cf. [48], [49] and Theorem 2.1). As regards a dynamic part, since in the problem in question the estimate (2.16) with the Gaussian kernel is bounded (see (2.5) and Assumptions 2.1-2.4), using this estimate we get in Stage 2 the convergence $\widehat{\gamma}_{N,M} \to \gamma$ in probability as $N, M \to \infty$ provided that $M \sim N^{5(1+\alpha)/2}$, or equivalently $N \sim M^{2/5(1+\alpha)}$, $\alpha > 0$ (cf. Theorem 2.3). For $\alpha = 1/2$ the asymptotic convergence rate is $\|\widehat{\gamma}_{N,M} - \gamma\| = O(N^{-1/2})$ in probability (Theorem 2.4). The behavior of the estimates $\widehat{c}_{N_0,M}$ and $\widehat{\gamma}_{N,M}$ for small and moderate number of data is demonstrated in the simulation example.

2.1.7 Simulation Study

The Hammerstein system from equation (2.5), with nonpolynomial static characteristic, was simulated for $m = 3$ and $s = 2$ taking $\phi(u_k) = (u_k, u_k^2, \sin u_k)^T$, $c = (2, 1, -20)^T$, $\gamma = (1, -1, 1)^T$, and $z_k = 2^{-1}z_{k-1} + \varepsilon_k$, i.e. $\{\omega_i = 2^{-i}\}_{i=0}^{\infty}$ (cf. (2.4)). Random input and white noise processes $\{u_k\}$ and $\{\varepsilon_k\}$ were generated according to the uniform distributions $u_k \sim U[-5; 5]$ and $\varepsilon_k \sim U[-\varepsilon_{max}; \varepsilon_{max}]$ (cf. Assumptions 2.1 and 2.4) where ε_{max} was changed as to give the noise-to-signal ratio $NSR = (z_{max}/v_{max}) \cdot 100\%$ equal to 1%, 5% and 10%, where $z_{max} = \varepsilon_{max} \sum_{i=0}^{\infty} 2^{-i} = 2\varepsilon_{max}$ is the noise magnitude (Assumption 2.4) and $v_{max} = w_{max} \sum_{i=0}^{2} |\gamma_i| = 3w_{max}$ with $w_{max} = \max_{u_k \in [-5;5]} |\phi^T(u_k)c|$ (cf. (2.3) and (2.5)) is the magnitude of the noiseless output signal; in our experiment $v_{max} = 165$. To estimate interactions in Stage 1, the nonparametric kernel estimate was applied, with the Gaussian kernel $K(u) = (1/\sqrt{2\pi})e^{-u^2/2}$ and the globally optimal bandwidth $h(M) = e(\varepsilon_{max})M^{-1/5}$, $e(\varepsilon_{max}) = (8.873 + 0.006\varepsilon_{max}^2)^{1/5}$, computed according to the rule recommended in [49] (see Section 8, p. 145 there). For the identification of the static element we assumed $N_0 = 4$ and the estimation points u_n in Stage 1 were chosen arbitrarily as $u_1 = -3.75$, $u_2 = -1.25$, $u_3 = 1.25$, and $u_4 = 3.75$. For the identification of the dynamic part we put $M = \lceil 0.5 \cdot N^{5(1+0.1)/2} \rceil = \lceil 0.5 \cdot N^{2.75} \rceil$, i.e. $N = \lceil (2M)^{1/2.75} \rceil = \lceil (2M)^{0.36} \rceil$. For each number M of data the experiment was repeated $P = 10$ times, and accuracy of the estimates $\widehat{c}_{N_0,M}$ and $\widehat{\gamma}_{N,M}$ was evaluated using the average relative estimation error $\delta_\theta(N, M) = \left[(1/P) \sum_{p=1}^{P} \left\| \widehat{\theta}_{N,M}^{(p)} - \theta \right\|_2^2 / \|\theta\|_2^2 \right] \cdot 100\%$, where $\widehat{\theta}_{N,M}^{(p)}$ is the estimate of $\theta \in \{c, \gamma\}$ obtained in the pth run, and $\|\cdot\|_2$ is the Euclidean vector norm. Exemplary results of two-stage identification of the nonlinear static characteristic for $M = 100$ and $M = 500$ measurement data and $NSR = 5\%$, along with the true characteristic and the 'data' points $\{(u_n, \widehat{w}_{n,M})\}_{n=1}^{N_0=4}$ (bold-faced) computed in Stage 1 by the kernel method, are visualized for a single trial in Fig. 2.2. The bold-faced points are the

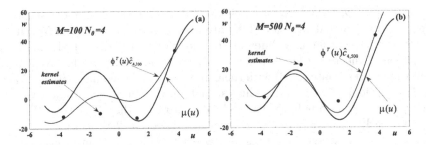

Fig. 2.2 True vs. estimated static characteristic and the 'data' points $(u_n, \widehat{w}_{n,M})$ (bold-faced) computed in Stage 1 by the kernel nonparametric method for (a) $M = 100$ (b) $M = 500$ measurement data

only outcome of the nonparametric estimation in the scheme. Thanks to the parametric prior knowledge of the characteristic we are able to derive from this small set of data the models $\hat{\mu}(u) = \phi^T(u)\widehat{c}_{N_0,M}$ $(N_0 = 4)$ of good visual quality in the whole range of inputs (particularly for $M = 500$), which is beyond the reach of any nonparametric method. The estimation error of vector c and γ is plotted in Fig. 2.3. The simulation results show convergence of the estimates $\widehat{c}_{N_0,M}$ and $\widehat{\gamma}_{N,M}$ with growing number M of measurement data. For comparison, the estimate of vector c was also computed by using the optimal two-stage least-squares SVD (2S-LS-SVD) algorithm [2], after obvious adaptation to Hammerstein systems. We assumed however incorrect prior knowledge of the system dynamics taking erroneously in the computations the dynamics order $s = 1$ instead of the true $s = 2$. Such a disagreement has of course no influence on the estimate $\widehat{c}_{N_0,M}$ in our method. The average relative estimation error for the 2S-LS-SVD algorithm and $NSR = 5\%$ is depicted in Fig. 2.3a with dashed line. As we see, the 2S-LS-SVD method gives in such circumstances biased estimate of c, with about 3% systematic error.

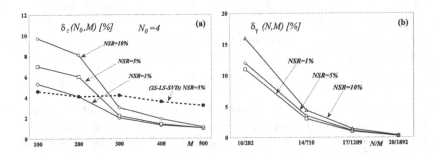

Fig. 2.3 The average relative estimation error vs. number of data for (a) vector c (dashed line: 2S-LS-SVD algorithm) (b) vector γ

2.1.8 Conclusions

Two stage parametric-nonparametric schemes for separate identification of parameters of a nonlinear static and linear FIR dynamic part of Hammerstein system have been proposed and examined. In each of the schemes only standard nonparametric regression and least squares computations are needed. Because of parametric prior knowledge the estimated models are provided in the explicit form. For a static part, the method works with nonpolynomial characteristics, thus extending the standard range of parametric methodology. The strength of the approach is complete decoupling of identification of the subsystems, computational simplicity, and robustness against lack or incorrectness of a priori knowledge of the noise model. Moreover, for each part of the system, the method is immune to possible errors in prior knowledge of companion subsystem. These advantages are achieved at the expense of a bit slower than the best possible convergence rate of the obtained parameter estimates, which is a consequence of the leading role of nonparametric regression estimation in the scheme and slower convergence of nonparametric methods.

2.2 Infinite Memory Hammerstein System

In this section the two-stage approach is generalized for the Hammerstein systems with the IIR dynamics (see [65] and [66]). Parameters of ARMA linear dynamic part of Hammerstein system (see [29]) are estimated by instrumental variables assuming poor a priori knowledge about static characteristic, the random input and random noise. Similarly as in Section 2.1, both subsystems are identified separately thanks to the fact that the unmeasurable interaction inputs and suitable instrumental variables are estimated in a preliminary step by the use of a nonparametric regression function estimation method. It is shown that the resulting estimates of linear subsystem parameters are consistent for both white and colored noise. The problem of generating optimal instruments is discussed and proper nonparametric method of computing the best instrumental variables is proposed. The analytical findings are then validated in Section 2.3.3, using numerical simulation results.

2.2.1 Introduction

In the proposed method, parametric and nonparametric approaches are also combined. Namely, the idea is to join the results obtained for the nonparametric identification of nonlinear characteristics in Hammerstein systems with IIR dynamics, in particular by using kernel regression methods ([48], [49]) with parametric knowledge of linear subsystem and standard results concerning least squares and instrumental variables (see, for instance, [128], [129], [150] and [154]), taking advantages of both. The algorithm is an extension of (2.14), where the combined parametric-nonparametric approach has been proposed for the identification of the impulse response of a FIR linear dynamics in Hammerstein system.

In Section 2.2.2 the identification problem is stated in detail and a priori assumptions on signals and subsystems of Hammerstein system are formulated. Stage 1 of the algorithm (nonparametric) consists in nonparametric estimation of interaction inputs $w_k = \mu(u_k)$ (Fig. 2.4) to cope with their inaccessibility for direct measurements. In Stage 2 (parametric), using the obtained estimates $\widehat{w}_k$ of w_k, we identify the linear dynamic subsystem. The instrumental variables based method is introduced and analyzed for estimation of parameters of $ARMA$ model. Both internal signals (inputs of the linear subsystem) and instruments are generated by the nonparametric regression estimation. The convergence rate is strictly proved and the problem of optimal, in the minimax sense, generation of instrumental variables is solved. In Section 2.3.3, the effectiveness of the approach under incomplete a priori knowledge of the static nonlinearity is also illustrated in simulation examples. The proofs are placed in the Appendices at the end of the book.

To summarize, a method of decomposing Hammerstein system identification task on fully independent partial identification problems of component subsystems, robust to the lack or inaccuracy of the prior knowledge of complementary subsystem, based on nonparametric estimation of interaction inputs, is developed. The identification algorithm of the linear dynamics of the $ARMA$-type (i.e. possessing infinite impulse response) with correlated output noise of arbitrary correlation structure, based on the concept of instrumental variables and nonparametric estimation of instruments is provided. Further, the convergence conditions and the asymptotic rate of convergence of adequate system parameter estimates are established. The problem of the optimum selection of instruments in the minimax sense is resolved and a suitable routine for estimating optimum instrumental variables is proposed. The efficiency of the proposed identification schemes, in particular their robustness to the incomplete or inaccurate knowledge of companion subsystems in the Hammerstein system is verified, and usability of the method, by means of computer simulations, is shown.

2.2.2   Statement of the Problem

IIR Hammerstein System

In this section we consider the discrete-time IIR Hammerstein system as in Fig. 2.4, where u_k, y_k and z_k are respectively the input, output, and noise at time k, and w_k is the interaction input not available for measurement, which is a typical limitation in the literature (see [13] for the discussion).

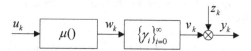

Fig. 2.4 Infinite memory Hammerstein system

Assumptions

We assume the following:

Assumption 2.6. *The input signal $\{u_k\}$ is for $k = ..., -1, 0, 1, ...$ an i.i.d. bounded random process $|u_k| \leqslant u_{\max}$, some $u_{\max} > 0$, and there exists a probability density of u_k, say $\nu(u)$.*

Assumption 2.7. *The nonlinear characteristic $\mu(u)$ is a Borel measurable bounded function on the interval $[-u_{\max}, u_{\max}]$, i.e.*

$$|\mu(u)| \leqslant w_{\max} \qquad (2.17)$$

where $w_{\max}$ is some unknown positive constant.

Assumption 2.8. *The linear dynamics is an asymptotically stable IIR filter*

$$v_k = \sum_{i=0}^{\infty} \gamma_i w_{k-i} \qquad (2.18)$$

with the infinite and unknown impulse response $\{\gamma_i\}_{i=0}^{\infty}$ (such that $\sum_{i=0}^{\infty} |\gamma_i| < \infty$).

Assumption 2.9. *The output noise $\{z_k\}$ is a random, arbitrarily correlated process, governed by the general equation*

$$z_k = \sum_{i=0}^{\infty} \omega_i \varepsilon_{k-i} \qquad (2.19)$$

where $\{\varepsilon_k\}$, $k = ..., -1, 0, 1, ...$, is a bounded stationary zero-mean white noise ($E\varepsilon_k = 0$, $|\varepsilon_k| \leqslant \varepsilon_{\max}$), independent of the input signal $\{u_k\}$, and $\{\omega_i\}_{i=0}^{\infty}$ is unknown; $\sum_{i=0}^{\infty} |\omega_i| < \infty$. Hence the noise $\{z_k\}$ is a stationary zero-mean and bounded process $|z_k| \leqslant z_{\max}$, where $z_{\max} = \varepsilon_{\max} \sum_{i=0}^{\infty} |\omega_i|$.

Assumption 2.10. *$\mu(u_0)$ is known at some point u_0 and $\gamma_0 = 1$.*

As it was explained in detail in [64], the input-output pair $(u_0, \mu(u_0))$ assumed to be known can refer to arbitrary $u_0 \in [-u_{\max}, u_{\max}]$, and hence we shall further assume, similarly as in Chapter 1, that $u_0 = 0$ and $\mu(0) = 0$, without loss of generality. The essence of Assumption 2.10 is explained in Remark 2.9.

As regards the parametric prior knowledge of the linear block, we additionally assume that:

Assumption 2.11. *The linear element (2.18) is an $ARMA(s,p)$ block, i.e. that it can be described in the parametric form by the following difference equation*

$$v_k = b_0 w_k + ... + b_s w_{k-s} + a_1 v_{k-1} + ... + a_p v_{k-p}, \qquad (2.20)$$

with known orders s, p and unknown parameters $b_0, b_1, ..., b_s$ and $a_1, a_2, ..., a_p$, or equivalently as $A(q^{-1})v_k = B(q^{-1})w_k$, where $B(q^{-1}) = b_0 + b_1 q^{-1} + ... + b_s q^{-s}$, $A(q^{-1}) = 1 - a_1 q^{-1} - ... - a_p q^{-p}$ and q^{-1} is a backward shift operator (see Fig. 2.5). Because of Assumption 2.10 it holds that $b_0 = 1$.

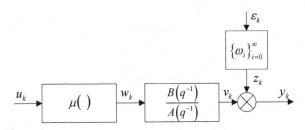

Fig. 2.5 Hammerstein system with $ARMA$ linear element

As it will be seen, when identifying dynamic subsystem, the parametric knowledge of the static element is not needed. The aim is to discover the true parameters $\theta = (b_0, b_1, ..., b_s, a_1, a_2, ..., a_p)^T$ of the linear block, using a set of input-output data $\{(u_k, y_k)\}$ collected from the whole system in an identification experiment.

2.2.3 Identification of ARMA Linear Dynamics by Nonparametric Instrumental Variables

Least Squares versus Instrumental Variables

Since $v_k = y_k - z_k$ (Fig. 2.5), thus the noisy dynamics output y_k can be expressed as (cf. (2.20))

$$y_k = \vartheta_k^T \theta + \bar{z}_k \qquad (2.21)$$

where $\theta = (b_0, b_1, ..., b_s, a_1, a_2, ..., a_p)^T$ is a vector of unknown true parameters of the linear dynamics,

$$\vartheta_k = (w_k, w_{k-1}, ..., w_{k-s}, y_{k-1}, y_{k-2}, ..., y_{k-p})^T$$

is a generalized input vector and

$$\bar{z}_k = z_k - a_1 z_{k-1} - \ldots - a_p z_{k-p}$$

is a proper, zero-mean and stationary resultant disturbance. For a set of N generalized input-output data $\{(\vartheta_k, y_k)\}$, we can write concisely

$$Y_N = \Theta_N \theta + Z_N$$

where

$$Y_N = (y_1, y_2, \ldots, y_N)^T,$$
$$\Theta_N = (\vartheta_1, \vartheta_2, \ldots, \vartheta_N)^T,$$
$$Z_N = (\bar{z}_1, \bar{z}_2, \ldots, \bar{z}_N)^T.$$

To this end we need in fact to have at disposal the input output data $\{(u_k, y_k)\}$ for $1 - p \leqslant k \leqslant N$, where p is the dynamics order.

Since the matrix Θ_N contains among others the regressors $y_{k-1}, y_{k-2}, \ldots$ $\ldots, y_{k-p}$ and the noise $\{\bar{z}_k\}$ is not white, the popular least squares estimate of θ, of the form

$$\hat{\theta}_N^{(LS)} = (\Theta_N^T \Theta_N)^{-1} \Theta_N^T Y_N \tag{2.22}$$

is obviously biased ([57], [129]). As is well known, we can overcome this weakness by using instrumental variables approach, yielding the estimate (see [128], [102])

$$\hat{\theta}_N^{(IV)} = (\Psi_N^T \Theta_N)^{-1} \Psi_N^T Y_N, \tag{2.23}$$

where Ψ_N is a matrix of properly selected instruments

$$\Psi_N = (\psi_1, \psi_2, \ldots, \psi_N)^T, \qquad \psi_k = (\psi_{k,1}, \psi_{k,2}, \ldots, \psi_{k,s+p+1})^T, \tag{2.24}$$

such that the following two properties hold
 (a) $\text{Plim}_{N \to \infty} \left(\frac{1}{N} \Psi_N^T \Theta_N \right)$ exists and is not singular
 (b) $\text{Plim}_{N \to \infty} \left(\frac{1}{N} \Psi_N^T Z_N \right) = 0$
for each $a_1, a_2, \ldots a_p$, with $Z_N = (\bar{z}_1, \bar{z}_2, \ldots, \bar{z}_N)^T$ and $\bar{z}_k = z_k - a_1 z_{k-1} - \ldots - a_p z_{k-p}$ as given above.

Remark 2.5. *Observe that*

$$Z_N = \left[Z_N^{(0)}, Z_N^{(1)}, \ldots, Z_N^{(p)} \right] \cdot \begin{bmatrix} 1 \\ -a_1 \\ \vdots \\ -a_p \end{bmatrix},$$

where

$$Z_N^{(r)} = (z_{1-r}, z_{2-r}, \ldots, z_{N-r})^T, \qquad r = 0, 1, \ldots, p, \tag{2.25}$$

and by virtue of (b) it holds that

$$\left[\frac{1}{N}\Psi_N^T Z_N^{(0)}, \frac{1}{N}\Psi_N^T Z_N^{(1)}, ..., \frac{1}{N}\Psi_N^T Z_N^{(p)}\right] \cdot \begin{bmatrix} 1 \\ -a_1 \\ \vdots \\ -a_p \end{bmatrix} \to 0$$

in probability as $N \to \infty$, for each $a_1, a_2, ..., a_p$. Hence, by lying $a_1 = a_2 = ... = a_p = 0$, we obtain in particular that

$$\frac{1}{N}\Psi_N^T Z_N^{(0)} \to 0, \text{ in probability, as } N \to \infty,$$

which implies in consequence that it holds

$$(-a_1)\frac{1}{N}\Psi_N^T Z_N^{(1)} + ... + (-a_p)\frac{1}{N}\Psi_N^T Z_N^{(p)} \to 0 \text{ in probability as } N \to \infty$$

for arbitrary $a_1, a_2, ..., a_p$. Consequently, by zeroing $a_j \neq a_r$ and putting $a_r = -1$, we ascertain that

$$\frac{1}{N}\Psi_N^T Z_N^{(r)} \to 0, \text{ in probability as } N \to \infty \qquad (2.26)$$

for each $r = 1, 2, ..., p$.

Under such conditions the estimation error

$$\Delta_N^{(IV)} = \widehat{\theta}_N^{(IV)} - \theta = \left(\frac{1}{N}\Psi_N^T \Theta_N\right)^{-1}\left(\frac{1}{N}\Psi_N^T Z_N\right) \qquad (2.27)$$

tends to zero (in probability) as $N \to \infty$, i.e. $\widehat{\theta}_N^{(IV)} \to \theta$ in probability as N grows large.

The conditions *(a)* and *(b)* require in fact the elements of Ψ_N be correlated with inputs and simultaneously not correlated with the noise $\{\overline{z}_k\}$. The simplest Ψ_N-generation techniques exploit directly former inputs of linear dynamics (see e.g. [131]), i.e. we take

$$\psi_{k,i} = w_{k-i+1}$$

which yields

$$\psi_k = (w_k, ..., w_{k-s}, w_{k-s-1}, ..., w_{k-s-p})^T \qquad (2.28)$$

Other, more sophisticated Ψ_N-generation methods are described in e.g. [127], [128], [129]. For clarity of exposition, we shall further confine to the estimate (2.23) with the instruments (2.28).

Nonparametric Instrumental Variables

As it was already pointed out, in the Hammerstein system the needed inputs $w_k, ..., w_{k-s-p}$ of the linear dynamics, $w_k = \mu(u_k)$, are not accessible for measurements, thus, precluding the direct use of the instrumental variables estimate (2.23), (2.28). According to our idea, to overcome this drawback we propose to compute appropriate estimates $\widehat{w}_{k,M}$ by a nonparametric technique (in Stage 1; see (2.12) and the Example in Section 2.3.2), similarly as for a static element, and use them instead of unknown w_k's, here for $1 - (s + p) \leqslant k \leqslant N$, obtaining the following plug in estimate

$$\widehat{\theta}_{N,M}^{(IV)} = (\widehat{\Psi}_{N,M}^T \widehat{\Theta}_{N,M})^{-1} \widehat{\Psi}_{N,M}^T Y_N \tag{2.29}$$

where

$$\widehat{\Theta}_{N,M} = (\widehat{\vartheta}_{1,M}, ..., \widehat{\vartheta}_{N,M})^T, \tag{2.30}$$

$$\widehat{\vartheta}_{k,M} = (\widehat{w}_{k,M}, ..., \widehat{w}_{k-s,M}, y_{k-1}, ..., y_{k-p})^T,$$

$$\widehat{\Psi}_{N,M} = (\widehat{\psi}_{1,M}, ..., \widehat{\psi}_{N,M})^T, \tag{2.31}$$

$$\widehat{\psi}_{k,M} = (\widehat{w}_{k,M}, ..., \widehat{w}_{k-s,M}, \widehat{w}_{k-s-1,M}, ..., \widehat{w}_{k-s-p,M})^T.$$

Remark 2.6. *The estimation points (treated as fixed points in Stage 1)*

$$u_1, ..., u_{1-s}, u_{1-s-1}, ..., u_{1-s-p}$$

$$u_2, ..., u_{2-s}, u_{2-s-1}, ..., u_{2-s-p}$$

$$\cdots\cdots\cdots\cdots\cdots\cdots\cdots$$

$$u_N, ..., u_{N-s}, u_{N-s-1}, ..., u_{N-s-p}$$

i.e.

$$u_{1-(s+p)}, u_{2-(s+p)}, ..., u_0; u_1, u_2, ..., u_N$$

for which $w_{k-r} = \mu(u_{k-r}) = \mu(u_{k-r}, c^*)$ have to be estimated are generated at random, by the input law, and $\{(u_{k-r}, y_{k-r}); k = 1, 2, ..., N$ and $r = 0, 1, ..., s + p\}$ induce the "data" $\{(w_{k-r}, y_{k-r}); k = 1, 2, ..., N$ and $r = 0, 1, ..., s + p\}$ required for the identification of linear dynamics. Further, $\{(u_k, y_k)\}_{k=1}^N$ (corresponding to $r = 0$) constitute first part of the data collection $\{(u_k, y_k)\}_{k=1}^M$, $M \gg N$, used for nonparametric estimation of w_{k-r}'s in Stage 1. The time instant index $k = 1$ is, by assumption, attributed to the beginning of recording the input-output data employed to this end.

Such a strategy yields a consistent estimate $\widehat{\theta}_{N,M}^{(IV)}$ of θ (see Theorems 2.5 and 2.6 below and proofs in Appendix A.8). It is a well known fact in the linear system identification theory that instrumental variables (2.28) guarantee fulfillment of *(b)* and of the nonsingularity condition *(a)* under whiteness of the signal $\{w_k\}$. In this context along with Assumption 2.6, the nonparametric estimation of interactions w_k $(= \mu(u_k))$ and next application of the obtained estimates in (2.29) seems to be justified and rather natural. In contrast to the method presented in [131], where the instruments were constructed by the specific linear filtering of the overall system input process $\{u_k\}$, our transformation of the inputs $\{u_k\}$ is more direct however nonlinear (see e.g. (2.54)).

Referring to (2.27) and the conditions *(a)* and *(b)* generally imposed on the instruments Ψ_N, for $\widehat{\Theta}_{N,M}$ and $\widehat{\Psi}_{N,M}$ given in (2.30), (2.31) and appearing in (2.29) we can show that the following theorems hold.

Theorem 2.5. *If the nonparametric estimate $\widehat{R}_M(u)$ is bounded, converges pointwise to the regression function $R(u)$ and at the estimation points $u \in \{0, u_{k-r};$ for $k = 1, 2, ..., N$ and $r = 0, 1, ..., s+p\}$ (see (2.50), (2.30), (2.31)) the error behaves like*

$$\left|\widehat{R}_M(u) - R(u)\right| = O(M^{-\tau}) \text{ in probability} \tag{2.32}$$

then
(a') $Plim_{M,N\to\infty}\left(\frac{1}{N}\widehat{\Psi}_{N,M}^T\widehat{\Theta}_{N,M}\right)$ does exist and is not singular
(b') $Plim_{M,N\to\infty}\left(\frac{1}{N}\widehat{\Psi}_{N,M}^T Z_N\right) = 0$
provided that $NM^{-\tau} \to 0$.

Proof. For the proof see Appendix A.8.   ■

Theorem 2.6. *Under assumptions of Theorem 2.5, for the estimate (2.29) with $\widehat{\Theta}_{N,M}$ and the instruments $\widehat{\Psi}_{N,M}$ as in (2.30) and (2.31) it holds that*

$$\widehat{\theta}_{N,M}^{(IV)} \to \theta \text{ in probability} \tag{2.33}$$

as $N, M \to \infty$, provided that $NM^{-\tau} \to 0$. Particularly, for $M \sim N^{(1+\alpha)/\tau}$, $\alpha > 0$, the asymptotic rate of convergence is

$$\left\|\widehat{\theta}_{N,M}^{(IV)} - \theta\right\| = O(N^{-\min(\frac{1}{2},\alpha)}) \text{ in probability} \tag{2.34}$$

Proof. Theorem may be proved in a similar manner as Theorems 2.3, and 2.4 with obvious substitutions, and hence the proof is here omitted.   ■

We observe that since in (2.32) in general we have $0 < \tau < 1/2$, hence to fulfil the conditions *(a')* and *(b')* far more input-output data, M, must be used in Stage 1 for nonparametric estimation of interactions $\{w_k\}$ than the inputs and instruments, N, for computation of the estimate $\widehat{\theta}_{N,M}^{(IV)}$ in Stage 2. This can be explained by the necessity of proper reduction of estimation errors of w_k and slower convergence of nonparametric estimates. For $M \sim N^{(1+\alpha)/\tau}$, any $\alpha > 0$, we get consistent estimate $\widehat{\theta}_{N,M}^{(IV)}$ (see (2.33)) and for $\alpha \geqslant 1/2$ the estimate attains (w.r.t. N) the best possible parametric rate of convergence $O(N^{-1/2})$ in probability (see (2.34)).

Selection of Optimal Instruments

Since the accuracy (2.27) of the *IV* method obviously depends on the employed instruments, the natural question is how to choose the instruments in an optimal manner. We shall shortly discuss this issue below and provide a suboptimal 'practical' rule for selection of the best (in some sense) instrumental variables for the Hammerstein system. The rule is based on nonparametric estimates of the interactions $\{w_k\}$ and of the noise-free outputs $\{v_k\}$ (see Fig. 2.5). Denote

$$\Gamma_N \triangleq \left(\frac{1}{N}\Psi_N^T\Theta_N\right)^{-1}\frac{1}{\sqrt{N}}\Psi_N^T$$

$$Z_N^* \triangleq \frac{\frac{1}{\sqrt{N}}Z_N}{\bar{z}_{\max}}$$

where $\bar{z}_{\max}$ is an upper bound on the absolute value of the noise $\{\bar{z}_k\}$ (see (2.21) and Assumption 2.9). The estimation error (2.27) can be clearly rewritten as

$$\Delta_N^{(IV)} = \bar{z}_{\max}\Gamma_N Z_N^* \tag{2.35}$$

Note that the Euclidean norm $\|\cdot\|_2$[1] of the normalized noise vector Z_N^* is now less or equal to 1:

$$\|Z_N^*\|_2 = \sqrt{\sum_{k=1}^N\left(\frac{\frac{1}{\sqrt{N}}\bar{z}_k}{\bar{z}_{\max}}\right)^2} = \sqrt{\frac{1}{N}\sum_{k=1}^N\left(\frac{\bar{z}_k}{\bar{z}_{\max}}\right)^2} \leq 1$$

Let us apply the following instruments' quality index

$$Q(\Psi_N) = \max_{\|Z_N^*\|_2 \leq 1}\left\|\Delta_N^{(IV)}(\Psi_N)\right\|_2^2 \tag{2.36}$$

We write here $\Delta_N^{(IV)}(\Psi_N)$ to emphasize the dependence of the error $\Delta_N^{(IV)}$ on the choice of Ψ_N, i.e. the instruments ψ_k. The following theorem can be proved.

Theorem 2.7. *For Hammerstein systems, the index $Q(\Psi_N)$ is asymptotically optimal for the instrumental matrix*

$$\Psi_N^* = (\psi_1^*, \psi_2^*, ..., \psi_N^*)^T, \tag{2.37}$$

$$with \ \psi_k^* = (w_k, w_{k-1}, ..., w_{k-s}, v_{k-1}, v_{k-2}, ..., v_{k-p})^T,$$

where $w_k, w_{k-1}, ..., w_{k-s}$ are interactions and $v_{k-1}, v_{k-2}, ..., v_{k-p}$ are noise-free outputs of the system (Fig. 2.5), i.e. for Ψ_N^* as in (2.37) and all other admissible choices of Ψ_N as in (2.24) it holds that

$$\lim_{N\to\infty} Q(\Psi_N^*) \leq \lim_{N\to\infty} Q(\Psi_N) \quad with \ probability \ 1 \tag{2.38}$$

Proof. For the proof see Appendix A.9. ∎

---

[1] From now on the Euclidean norm will be denoted by $\|\cdot\|_2$ to avoid ambiguity.

Theorem 2.7 is only of theoretical value because of inaccessibility in the system of $\{w_k\}$ and $\{v_k\}$. However it provides a guideline concerning the best choice of instruments, which can be used as a starting point for setting up a 'practical' routine for synthesis of the instruments. Namely, using a nonparametric technique (particularly, nonparametric estimates $\widehat{w}_k$ of w_k) and taking account of the form of the optimal instruments (2.37), we can propose a natural scheme, where instrumental variable vectors are computed as

$$\widehat{\psi}^*_{k,M} = (\widehat{w}_{k,M}, \widehat{w}_{k-1,M}, ..., \widehat{w}_{k-s,M}, \widehat{v}_{k-1,M}, \widehat{v}_{k-2,M}, ..., \widehat{v}_{k-p,M})^T, \quad (2.39)$$

where $\widehat{w}_{k,M}$'s are nonparametric estimates (2.50) of w_k's and $\widehat{v}_{k,M}$'s are nonparametric estimates of v_k's calculated as

$$\widehat{v}_{k,M} = \sum_{i=0}^{F} \widehat{\gamma}_{i,M} \widehat{w}_{k-i,M}$$

with (see [48])

$$\widehat{\gamma}_{i,M} = \widehat{\varkappa}_{i,M}/\widehat{\varkappa}_{0,M}, \qquad \widehat{\varkappa}_{i,M} = \frac{1}{M} \sum_{k=1}^{M-i} (y_{k+i} - \overline{y})(u_k - \overline{u}),$$

$$\overline{y} = \frac{1}{M} \sum_{k=1}^{M} y_k, \qquad \overline{u} = \frac{1}{M} \sum_{k=1}^{M} u_k,$$

and F being a chosen "cut-off level" of the infinite length impulse response $\{\gamma_i\}$ of the linear dynamics in the Hammerstein system. The simulations presented in Section 2.3.3 confirm efficiency of such a scheme, however comprehensive theoretical analysis, in particular examination of the influence of the selection of the F value on the asymptotic behavior of the estimate $\widehat{\theta}^{(IV)}_{N,M}$ with $\widehat{\Psi}_{N,M} = \widehat{\Psi}^*_{N,M}$ $(= (\widehat{\psi}^*_{1,M}, \widehat{\psi}^*_{2,M}, ..., \widehat{\psi}^*_{N,M})^T)$ is an open question and is left for future research.

2.2.4 Conclusions

The two-stage parametric-nonparametric approach for the identification of IIR Hammerstein systems by instrumental variables method reveals the following noteworthy properties: 1) small a priori knowledge about the random signals needed for the identification scheme to work and generality of the identification framework, as any particular model of the input and noise is not assumed, 2) broad applicability, as vast class of non-linear characteristics and $ARMA$-type linear dynamics (with infinite impulse response) is admitted, 3) full decomposition of the Hammerstein system identification task, as identification of subsystems is performed in fact in a completely decentralized (local) manner, 4) robustness to the partial inaccuracy of a priori knowledge, as successful identification of one system component can be performed in spite of imprecise parametric information of the other, 5) computational simplicity, as parametric identification of subsystems can be performed independently (locally) by using standard identification routines with well elaborated software, and nonparametric stage needs only elementary computations. The disadvantage is that some deterioration of convergence speed of the parameter estimates, in comparison with the best possible parametric rate of convergence, can occur in the method. This is caused by slower convergence of nonparametric techniques employed in the first (preliminary) stage in the approach. General conditions are provided for obtaining successful symbiosis of parametric and nonparametric identification methods, i.e. of getting hybrid parametric-nonparametric identification algorithms which guarantee achievement of consistent and effective estimates (cf. Theorems 2.5-2.7) in a convenient way and at no great expense.

2.3 Hammerstein System with Hard Nonlinearity

In Section 2.2, the FIR linear block of Hammerstein system (2.3) was generalized for the IIR ARMA filter (see (2.18) and (2.20)). Here, we generalize the parametric form of the static nonlinear block (2.1). Problem decomposition by application of mixed parametric-nonparametric approach allows to solve the identification task for a broad class of nonlinear characteristics including functions, which are *not linear in the parameters*. The idea presented below is an extension of the two-stage estimate described in Section 2.1, where the combined parametric-nonparametric algorithm has been proposed for the identification of parameters appearing linearly in the static nonlinear element. In Stage 2 of the procedure, using the nonparametric estimates $\widehat{w}_n$ of w_n, computed in Stage 1, we identify parameters of nonlinearity by minimization of appropriate empirical quadratic criterion function. It is shown in particular, that the quadratic criterion function is bounded from the above and from the below by two paraboloids. This fact is further exploited for the proving consistency of the parameter estimate of the nonlinearity. The parameter identifiability conditions are discussed and the consistency of the resulting parameter estimate of the nonlinearity is shown. The asymptotic rate of convergence is also established. We emphasize that the knowledge of the impulse response of the linear dynamic block is still nonparametric.

2.3.1 Parametric Prior Knowledge of the Nonlinear Characteristic

In this point, additional assumptions (with respect to Section 2.2) specifies parametric a priori knowledge of $\mu()$ which is available in our identification task. We suppose the following:

Assumption 2.12. *The form of a static nonlinearity $\mu()$ is known up to the parameters, i.e. we are given the function $\mu(u, c)$ and the set C of admissible parameters (it is the pair $(\mu(u, c), C)$) such that $\mu(u, c^*) = \mu(u)$ (Fig. 2.6), where $c^* = (c_1^*, c_2^*, ..., c_m^*)^T \in C$ is a vector of the unknown true parameters of the nonlinearity.*

Assumption 2.13. *The function $\mu(u, c)$ is continuous and differentiable in the set C, and the gradient $\nabla_c \mu(u, c)$ is bounded, i.e.*

$$\|\nabla_c \mu(u, c)\| \leqslant G_{\max} < \infty, \qquad c \in C$$

for each $u \in [-u_{\max}, u_{\max}]$.

Assumption 2.14. *The admissible set of parameters C is bounded and small enough (cf. Remark 2.7 below), and can be considered as a neighbourhood of each admissible parameter vector $c \in C$, and in particular, of the true parameter vector c^*, i.e. $C \equiv \mathcal{O}(c^*)$.*

Assumption 2.15. *Each parameter $c \in C$, and in particular the true parameter vector $c^* \in C$ is identifiable in the set C, i.e. there exists a sequence of inputs (design points) $\overline{u}_1, \overline{u}_2, ..., \overline{u}_{N_0}$, independent of c, such that*

$$\mu(\overline{u}_n, c) = \mu(\overline{u}_n, c^*), \ n = 1, 2, ..., N_0 \Rightarrow c = c^* \tag{2.40}$$

for each pair $(c, c^*) \in C \times C$.

Remarks below present some comments and motivations concerning above Assumptions.

Remark 2.7. *Assumptions 2.12 and 2.13 constitute general (and typical) parametric a priori information about the real static nonlinearity $\mu(u)$. However many practical problems involve nonlinear effects that are prone to a high amount of detailed structure and hence are difficult to describe parametrically by a single model as in Assumption 2.12. In circumstances where yet single and smooth (Assumption 2.13) model is planned to be used to handle nonlinearity, it should be kept in mind that in the real world its use is possible only locally, both with respect to the inputs (Assumption 2.3) and the parameters (Assumption 2.14), and can require a good deal of expertise. For comprehensive discussion of this and related issues we refer the reader to [114].*

Remark 2.8. *As regards Assumption 2.15 for $c \in C$, where Assumptions 2.13 and 2.14 allow a one-term Taylor expansion of the form*

$$\mu(\overline{u}_n, c) = \mu(\overline{u}_n, c^*) + \nabla_c^T \mu(\overline{u}_n, c^*)(c - c^*) + o\left(\|c - c^*\|\right)$$

and $o\left(\|c - c^*\|\right)$ is negligible, we have factually

$$\mu(\overline{u}_n, c) - \mu(\overline{u}_n, c^*) = \nabla_c^T \mu(\overline{u}_n, c^*)(c - c^*), \tag{2.41}$$

hence the identifiability condition (2.40) can be rewritten as

$$\nabla_c^T \mu(\overline{u}_n, c^*)(c - c^*) = 0, \ n = 1, 2, ..., N_0 \Rightarrow c = c^*$$

or that

$$J(\overline{u}_1, \overline{u}_2, ..., \overline{u}_{N_0}; c^*)(c - c^*) = 0$$

implies $c = c^*$, *where*

$$J(\overline{u}_1, \overline{u}_2, ..., \overline{u}_{N_0}; c^*) = [\nabla_c \mu(\overline{u}_1, c^*), \nabla_c \mu(\overline{u}_2, c^*), ..., \nabla_c \mu(\overline{u}_{N_0}, c^*)]^T$$

is the Jacobian matrix. This means that – from the viewpoint of the true (but unfortunately unknown) parameter vector c^* – the input sequence $\{\overline{u}_n; n = 1, 2, ..., N_0\}$ in (2.40) should be selected such that

$$\text{rank} J(\overline{u}_1, \overline{u}_2, ..., \overline{u}_{N_0}; c^*) = \dim c, \qquad (2.42)$$

i.e. $N_0 \geqslant \dim c$ (the necessary condition) and

$$\det J^T(\overline{u}_1, \overline{u}_2, ..., \overline{u}_{N_0}; c^*) J(\overline{u}_1, \overline{u}_2, ..., \overline{u}_{N_0}; c^*) > 0.$$

To get in turn independent of c^* and of any $c \in C$ (universal) choice of proper $\overline{u}_n$'s one should need that

$$\inf_{c \in C} \left\{ \det J^T(\overline{u}_1, \overline{u}_2, ..., \overline{u}_{N_0}; c) J(\overline{u}_1, \overline{u}_2, ..., \overline{u}_{N_0}; c) \right\} > 0. \qquad (2.43)$$

In the particular case of linear-in-the-parameters static characteristics

$$\mu(u, c) = \phi^T(u)c$$

discussed in [64], where $\phi(u) = (f_1(u), f_2(u), ..., f_m(u))^T$ is a vector of linearly independent basis functions, these requirements take the form (compare Remark 2 in [64])

$$\text{rank}(\phi(\overline{u}_1), \phi(\overline{u}_2), ..., \phi(\overline{u}_{N_0}))^T = \dim c,$$

or

$$\det \Phi_{N_0}^T(\overline{u}_1, \overline{u}_2, ..., \overline{u}_{N_0}) \Phi_{N_0}(\overline{u}_1, \overline{u}_2, ..., \overline{u}_{N_0}) > 0,$$

where $\Phi_{N_0}(\overline{u}_1, \overline{u}_2, ..., \overline{u}_{N_0}) = (\phi(\overline{u}_1), \phi(\overline{u}_2), ..., \phi(\overline{u}_{N_0}))^T$.

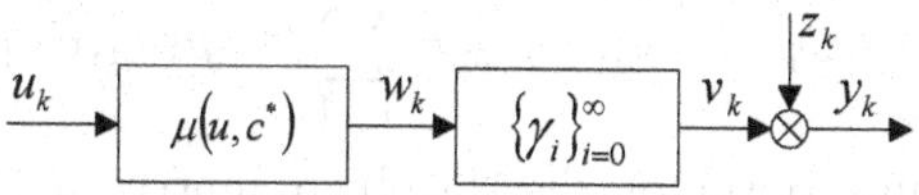

Fig. 2.6 Hammerstein system with parametric nonlinearity $\mu(u) = \mu(u, c^*)$

Conversely, we shall see that Assumption 2.11 from Section 2.2 may be weakened to Assumption 2.8 during identification process of the static part.

2.3.2 Estimation of the Nonlinearity Parameters

Let $\overline{u}_1, \overline{u}_2, ..., \overline{u}_{N_0}$ be such that the unknown c^* is identifiable (see Remark 2.8). Denote

$$W_{N_0} = (w_1, w_2, ..., w_{N_0})^T \tag{2.44}$$

where $w_n = \mu(\overline{u}_n, c^*)$ for $n = 1, 2, ..., N_0$ and introduce the vector

$$\overline{\mu}_{N_0}(c) = [\mu(\overline{u}_1, c), \mu(\overline{u}_2, c), ...\mu(\overline{u}_{N_0}, c)]^T$$

Obviously, the square of the Euclidean norm of the difference between $\overline{\mu}_{N_0}(c)$ and W_{N_0}, i.e., the index

$$Q_{N_0}(c) = \left\| \overline{\mu}_{N_0}(c) - W_{N_0} \right\|^2$$

takes minimum value only for the vector of true parameters c^* of the nonlinearity, i.e., the minimizer

$$c^* = \arg\min \left\| \overline{\mu}_{N_0}(c) - W_{N_0} \right\|^2$$

is unique and hence the identification routine may be based on the minimization of the residual sum of squares

$$Q_{N_0}(c) = \sum_{n=1}^{N_0} [w_n - \mu(\overline{u}_n, c)]^2; \qquad c \in C. \tag{2.45}$$

Because of $w_n = \mu(\overline{u}_n, c^*)$ and due to (2.41) in Remark 2.8, we have

$$Q_{N_0}(c) = \sum_{n=1}^{N_0} \left| \nabla_c^T \mu(\overline{u}_n, c^*)(c - c^*) \right|^2,$$

and by the Schwartz inequality we ascertain that

$$Q_{N_0}(c) \leqslant \left\| J(\overline{u}_1, \overline{u}_2, ..., \overline{u}_{N_0}; c^*) \right\|_F^2 \cdot \left\| c - c^* \right\|^2,$$

where

$$\left\| J(\overline{u}_1, \overline{u}_2, ..., \overline{u}_{N_0}; c^*) \right\|_F^2 = \sum_{n=1}^{N_0} \left\| \nabla_c \mu(\overline{u}_n, c^*) \right\|^2$$

is squared Frobenius norm of the Jacobian matrix $J(\overline{u}_1, \overline{u}_2, ..., \overline{u}_{N_0}; c^*)$. Hence, by virtue of Assumption 2.13 we get

$$\left\| J(\overline{u}_1, \overline{u}_2, ..., \overline{u}_{N_0}; c^*) \right\|_F^2 \leqslant N_0 G_{\max}^2 \triangleq D,$$

which yields that

$$Q_{N_0}(c) \leqslant D \cdot \|c - c^*\|^2, \qquad \forall c \in C. \tag{2.46}$$

On the other hand, since

$$Q_{N_0}(c) = (c - c^*)^T J^T(\overline{u}_1, \overline{u}_2, ..., \overline{u}_{N_0}; c^*) J(\overline{u}_1, \overline{u}_2, ..., \overline{u}_{N_0}; c^*)(c - c^*)$$

and for the selected $\overline{u}_1, \overline{u}_2, ..., \overline{u}_{N_0}$ the matrix

$$J^T(\overline{u}_1, \overline{u}_2, ..., \overline{u}_{N_0}; c^*) J(\overline{u}_1, \overline{u}_2, ..., \overline{u}_{N_0}; c^*)$$

is symmetric and positive-definite (by fulfillment of the identifiability condition for the selected input series $\overline{u}_1, \overline{u}_2, ..., \overline{u}_{N_0}$; see Remark 2.8), there exists a $\delta > 0$ such that

$$Q_{N_0}(c) \geqslant \delta \cdot \|c - c^*\|^2, \qquad \forall c \in C. \tag{2.47}$$

This follows from the observation that

$$\delta \cdot \|c - c^*\|^2 = (c - c^*)^T \delta I (c - c^*),$$

the fact that all eigenvalues of the matrix $J^T(...)J(...)$, say $\{\lambda_i\}_{i=1}^m$, are positive, $\lambda_i > 0$ for $i = 1, 2, ..., m$ (see e.g. the property 6.O, page 335 in [134]), and the fact that eigenvalues of the matrix

$$J^T(\overline{u}_1, \overline{u}_2, ..., \overline{u}_{N_0}; c^*) J(\overline{u}_1, \overline{u}_2, ..., \overline{u}_{N_0}; c^*) - \delta I$$

are $\{\lambda_i - \delta\}_{i=1}^m$ (see Lemma B.4 in Appendix B.3), whence it suffices to take some $0 < \delta \leqslant \min\{\lambda_i\}_{i=1}^m$ to assure (2.47), or, in other words, positive semidefiniteness of the matrix $J^T(\overline{u}_1, \overline{u}_2, ..., \overline{u}_{N_0}; c^*) J(\overline{u}_1, \overline{u}_2, ..., \overline{u}_{N_0}; c^*) - \delta I$. The bounds (2.46) and (2.47) yield together that in the parameter set C it holds that

$$\delta \cdot \|c - c^*\|^2 \leqslant Q_{N_0}(c) \leqslant D \cdot \|c - c^*\|^2 \tag{2.48}$$

or equivalently (due to $Q_{N_0}(c^*) = 0$) that

$$\delta \cdot \|c - c^*\|^2 \leqslant Q_{N_0}(c) - Q_{N_0}(c^*) \leqslant D \cdot \|c - c^*\|^2 \tag{2.49}$$

This shows that in the case under discussion the loss function $Q_{N_0}(c)$ with $c \in C$, computed for the input sequence $\overline{u}_1, \overline{u}_2, ..., \overline{u}_{N_0}$, is located in a region bounded by two paraboloids originated at c^*, and certifies identifiability of the parameter vector c^* for the selected input points $\{\overline{u}_n; n = 1, 2, ..., N_0\}$ (as $Q_{N_0}(c) = 0$ implies $c = c^*$). We stress that in our consideration $Q_{N_0}(c)$ in itself does not need to be a convex function in the set C.

Because of inaccessibility of interactions $\{w_n\}_{n=1}^{N_0}$ appearing in (2.45), the direct minimization of $Q_{N_0}(c)$ w.r.t. c is not possible. Instead, an indirect two stage procedure can be proposed where, in Stage 1, a nonparametric estimation of w_n's is carried out yielding the estimates $\widehat{w}_n$, and next, in Stage 2, a parametric minimization of $Q_{N_0}(c)$ is completed employing the obtained $\widehat{w}_n$ instead of w_n. The arising identification scheme is then the following.

Select $\bar{u}_1, \bar{u}_2, ..., \bar{u}_{N_0}$ so that the unknown parameter vector c^* be identifiable, i.e. due to the selection rule (2.43) in Remark 2.8.

Stage 1 (nonparametric): On the basis of M input-output measurement data $\{(u_k, y_k)\}_{k=1}^{M}$, for the selected N_0 input points $\{\bar{u}_n;\ n = 1, 2, ..., N_0\}$ estimate the corresponding interactions $\{w_n = \mu(\bar{u}_n, c^*);\ n = 1, 2, ..., N_0\}$ as

$$\widehat{w}_{n,M} = \widehat{R}_M(\bar{u}_n) - \widehat{R}_M(0), \qquad (2.50)$$

where $\widehat{R}_M(u)$ is a nonparametric estimate of the regression function $R(u) = E[y_k | u_k = u]$.

Stage 2 (parametric): Plug in the estimates $\widehat{w}_{n,M}$ obtained in Stage 1 to the loss function (2.45) in place of w_n and minimize the quality index

$$\widehat{Q}_{N_0,M}(c) = \sum_{n=1}^{N_0} [\widehat{w}_{n,M} - \mu(\bar{u}_n, c)]^2 \qquad (2.51)$$

getting the solution $\widehat{c}_{N_0,M}$. Take the computed $\widehat{c}_{N_0,M}$ as the estimate of c^*.

The idea behind Stage 1 originates from the fact that under Assumptions 2.6÷2.10 it holds that (see, e.g., [48])

$$R(u) = \gamma_0 \mu(u) + d$$

where $d = E\mu(u_1) \sum_{i=1}^{\infty} \gamma_i$ and $\mu()$ is the Hammerstein system nonlinearity, and under Assumption 2.10 along with the fact that by assumption $\mu(0) = 0$ we further get that

$$\mu(u, c^*) = R(u) - R(0).$$

Remark 2.9. *In fact, Assumption 2.10 can be generalized as in general fashion it holds that*

$$E\{y_{k+l} | u_k = u\} = \gamma_l \mu(u) + \delta_l,$$

where $\delta_l = E\mu(u_1) \cdot \sum_{i \neq l} \gamma_i$, i.e. the shifted regression of the output on the input in Hammerstein system is a scaled and properly shifted version of the nonlinear characteristic $\mu()$, i.e. factually we can estimate $\mu(u)$ for each timelag l between output and input, for which $\gamma_l \neq 0$. Certainly, the best choice of l is for the $|\gamma_l|$ being maximal. For further discussion we refer to [91] and [92].

Such a routine is consistent, i.e., provides the estimate $\widehat{c}_{N_0,M}$ which converges (as $M \to \infty$) to the true parameter vector c^* provided that $\widehat{R}_M(u)$ is a consistent estimate of the regression function $R(u)$. The following theorem refers to this property.

Theorem 2.8. *Assume that the computed $\widehat{c}_{N_0,M}$ is unique and for each M, $c_{N_0,M} \in C$. If in Stage 1 (nonparametric), for some $\tau > 0$, it holds that*

$$\widehat{R}_M(\bar{u}_n) = R(\bar{u}_n) + O(M^{-\tau}) \text{ in probability} \tag{2.52}$$

as $M \to \infty$ for $n = 1, 2, ..., N_0$ and for $\bar{u}_n = 0$ then

$$\widehat{c}_{N_0,M} = c^* + O(M^{-\tau}) \text{ in probability} \tag{2.53}$$

as $M \to \infty$.

Proof. For the proof see Appendix A.10. ∎

The theorem says that under our assumptions both convergence and the guaranteed rate of convergence of the nonparametric estimate $\widehat{R}_M(u)$ are conveyed on the estimate $\widehat{c}_{N_0,M}$ of the parameter vector c^*, i.e. the estimate $\widehat{c}_{N_0,M}$ converges to c^* in the same sense and with the same guaranteed speed as $\widehat{R}_M(u)$ to $R(u)$. The proof shows that each kind of probabilistic convergence of $\widehat{c}_{N_0,M}$ could potentially be considered in Theorem 2.8. We confined ourselves to the convergence in probability as such particular type of convergence has been widely examined in the literature concerning non-parametric estimation of nonlinearities (regression functions) for Hammerstein systems (see [40], [48], [49]). As is well known, the nonparametric rate of convergence $O(M^{-\tau})$ appearing in (2.52) is usually of slower order than typical parametric rate of convergence $O(M^{-1/2})$ in probability, i.e. as a rule $0 < \tau < 1/2$ (cf. e.g. [59]). Thus in our method, we can loose a little (for $\tau \approx 1/2$) the convergence speed of the estimate of c^*, but instead the successful recovering of c^* can be performed without prior knowledge (and hence even under false knowledge) of a companion dynamic part of the system. Moreover, for smooth nonlinearities $\mu(u) = \mu(u, c^*)$ the convergence rate can be made arbitrarily close to $O(M^{-1/2})$ by applying proper nonparametric regression function estimates $\widehat{R}_M(u)$ (cf. e.g. [40], [49], [69]).

The class of consistent nonparametric regression estimates $\widehat{R}_M(u)$, which can be employed in Stage 1, and were elaborated up to now in the system identification literature for Hammerstein systems, encompasses kernel and orthogonal series estimates, including wavelet estimates possessing excellent adaptation properties and being able to attain the best possible nonparametric rate of convergence in the sense of Stone (cf. [69], [132], and [133]). An excellent introduction to the broad field of nonparametric regression provides [136]. For a general treatment of nonparametric regression function estimation methods we send the reader to, e.g. [30], [59] and [149].

As a simple example from a broad variety of accessible nonparametric regression function estimates we present below the kernel estimate, very easy for computation.

Example

Kernel regression estimate (studied in [48], [49]) has the form

$$\widehat{R}_M(u) = \frac{\sum_{k=1}^{M} y_k K(\frac{u-u_k}{h(M)})}{\sum_{k=1}^{M} K(\frac{u-u_k}{h(M)})} = \sum_{k=1}^{M} \left[\frac{K(\frac{u-u_k}{h(M)})}{\sum_{k=1}^{M} K(\frac{u-u_k}{h(M)})} \right] \cdot y_k \qquad (2.54)$$

where $K(u)$ is a kernel (weighting) function and $h(M)$ is a bandwidth parameter controlling the range of data used for estimating a regression function $R(u)$ at a given point u. Standard examples are $K(u) = I_{[-0.5,0.5]}(u)$, $(1-|u|)I_{[-1,1]}(u)$ or $(1/\sqrt{2\pi})\,e^{-u^2/2}$ and $h(M) = const \cdot M^{-\alpha}$ with $0 < \alpha < 1$. Owing to the convergence results provided in [149], we find out that for each of the above kernels $K(u)$ it holds that $\widehat{R}_M(u) \to R(u)$ in probability as $M \to \infty$, and that the convergence takes place at every point $u \in Cont(\mu, \nu)$, the set of continuity points of $\mu(u)$ and $\nu(u)$, at which $\nu(u) > 0$ where $\nu(u)$ is a probability density of the system input (assumed to exist, cf. Assumption 2.6). Applying in particular the Gaussian kernel $K(u) = (1/\sqrt{2\pi})\,e^{-u^2/2}$ and taking, according to the recommendation in [49], $h(M) \sim M^{-1/5}$ we attain in Stage 1 the convergence rate $\left|\widehat{R}_M(u) - R(u)\right| = O(M^{-2/5})$ in probability very close to $O(M^{-1/2})$, provided that $\mu(u)$ and $\nu(u)$ are at the selected estimation points $u \in \{0, \overline{u}_n; n = 1, 2, ..., N_0\}$ (see (2.50) and (2.52)) two times or more continuously differentiable functions and $\nu(u) > 0$ there (cf. [48], [49]).

As regards the parametric optimization task (unconstrained nonlinear least squares) to be solved in Stage 2 we can use to this end for instance the standard Levenberg-Marquardt method. Minimization of $\widehat{Q}_{N_0,M}(c)$ in Stage 2 of the identification procedure is then performed with the use of the following iterative routine:

$$\widehat{c}_{N_0,M}^{(i+1)} = \widehat{c}_{N_0,M}^{(i)} - \left[J^T(\widehat{c}_{N_0,M}^{(i)}) J(\widehat{c}_{N_0,M}^{(i)}) + \lambda_i I \right]^{-1} J^T(\widehat{c}_{N_0,M}^{(i)}) r(\widehat{c}_{N_0,M}^{(i)})$$

where $J(c) = J(\overline{u}_1, \overline{u}_2, ..., \overline{u}_{N_0}; c^*)$ is the Jacobian matrix of the form $J(c)[n,j] = \frac{\partial r_n(c)}{\partial c_j}$ $(n = 1, 2, ..., N_0; j = 1, 2, ..., m)$, where $r_n(c) = \mu(\overline{u}_n, c) - \widehat{w}_{n,M}$ and $r(c) = (r_1(c), r_2(c), ..., r_{N_0}(c))^T$. The λ_i's are weighting coefficients (continuation parameters) modified according to the rule

$$\lambda_{i+1} = \begin{cases} \lambda_i \cdot \nu, \text{ if } \widehat{Q}_{N_0,M}(\widehat{c}_{N_0,M}^{(i+1)}) \geqslant \widehat{Q}_{N_0,M}(\widehat{c}_{N_0,M}^{(i)}) \\ \lambda_i/\nu, \text{ otherwise} \end{cases},$$

where $\nu > 1$ (see [86], [110] and [123] for various implementations and discussion of Levenberg-Marquardt method).

2.3.3 Simulation Examples

Identification of the Hammerstein System by the Use of the Levenberg-Marquardt Method and the Kernel Nonparametric Instrumental Variables

The Hammerstein system composed of the following blocks was simulated:
a) static nonlinearity (in a parametric form)

$$\mu(u, c) = c_1 u + c_2 u^2 + c_3 u^3 + e^{c_4 u} \stackrel{c=c^*}{=} u + u^2 - u^3 + e^u$$

with the unknown true parameter vector $c = c^* = (1, 1, -1, 1)^T$, and
b) linear dynamics (in a parametric form)

$$v_k = b_0 w_k + b_1 w_{k-1} + a_1 v_{k-1} + a_2 v_{k-2}$$

governed by the equation

$$v_k = w_k + w_{k-1} + 0.5 v_{k-1} + 0.25 v_{k-2}$$

with the unknown vector of true parameters $\theta = (1, 1, 0.5, 0.25)^T$.

The system was excited by an i.i.d. uniformly distributed input signal $u_k \sim U(-3, 3)$, and disturbed by the correlated ARMA noise $z_k = 0.7 z_{k-1} + \varepsilon_k + \varepsilon_{k-1}$, where $\varepsilon_k \sim U(-\delta, \delta)$ and magnitude δ was changed and set as $\delta = 0.1$, 0.2 and 0.3.

In the identification step of the nonlinear part, it was also assumed we know in advance that $\mu(0) = 1$, i.e., that the value of static characteristic $\mu(u, c^*) = \mu(u)$ is known at the point $u_0 = 0$ (see Assumption 2.10). Particularly to the relation $R(u) = \mu(u) + d$ (for $\gamma_0 = 1$), this requires a slight modification of the w_n's estimation routine (2.50), namely its revision to the form

$$\widehat{w}_{n,M} = \widehat{R}_M(\bar{u}_n) - \widehat{R}_M(0) + \mu(0),$$

i.e., to estimate the w_n's we actually need to use in Stage 1 the rule $\widehat{w}_{n,M} = \widehat{R}_M(\bar{u}_n) - \widehat{R}_M(0) + 1$ instead of (2.50).

Since $\dim c = 4$, in Stage 1 of our procedure, according to Remark 2.8, we choose for simplicity $N_0 = 4$. The kernel estimate described in the above *Example*, with the compactly supported window kernel

$$K(x) = \begin{cases} 1/2, & |x| < 1 \\ 0, & |x| \geqslant 1 \end{cases},$$

and $h(M) = 0.44 M^{-1/5}$ was computed, according to the rule recommended in [49] (see Section 8, p. 145). In turn, in Stage 2, Levenberg-Marquardt methodology has been applied for minimization of the appropriate loss function of the form (cf. (2.51))

$$\widehat{Q}_{N_0,M}(c) = \sum_{n=1}^{4} \left[c_1 \overline{u}_n + c_2 \overline{u}_n^2 + c_3 \overline{u}_n^3 + e^{c_4 \overline{u}_n} - \widehat{w}_{n,M} \right]^2 = \sum_{n=1}^{4} r_n^2(c)$$

being the empirical version of the index (2.45), with the Jacobian matrix $J(\overline{u}_1, \overline{u}_2, ..., \overline{u}_{N_0}; c)$ (see Remark 2.8), denoted for shortness as $J(c)$, of the form

$$J(c) = \begin{bmatrix} \frac{\partial \mu(\overline{u}_1, c)}{\partial c_1} & \frac{\partial \mu(\overline{u}_1, c)}{\partial c_2} & \frac{\partial \mu(\overline{u}_1, c)}{\partial c_3} & \frac{\partial \mu(\overline{u}_1, c)}{\partial c_4} \\ \frac{\partial \mu(\overline{u}_2, c)}{\partial c_1} & \frac{\partial \mu(\overline{u}_2, c)}{\partial c_2} & \frac{\partial \mu(\overline{u}_2, c)}{\partial c_3} & \frac{\partial \mu(\overline{u}_2, c)}{\partial c_4} \\ \frac{\partial \mu(\overline{u}_3, c)}{\partial c_1} & \frac{\partial \mu(\overline{u}_3, c)}{\partial c_2} & \frac{\partial \mu(\overline{u}_3, c)}{\partial c_3} & \frac{\partial \mu(\overline{u}_3, c)}{\partial c_4} \\ \frac{\partial \mu(\overline{u}_4, c)}{\partial c_1} & \frac{\partial \mu(\overline{u}_4, c)}{\partial c_2} & \frac{\partial \mu(\overline{u}_4, c)}{\partial c_3} & \frac{\partial \mu(\overline{u}_4, c)}{\partial c_4} \end{bmatrix} =$$

$$= \begin{bmatrix} \frac{\partial r_1}{\partial c_1} & \frac{\partial r_1}{\partial c_2} & \frac{\partial r_1}{\partial c_3} & \frac{\partial r_1}{\partial c_4} \\ \frac{\partial r_2}{\partial c_1} & \frac{\partial r_2}{\partial c_2} & \frac{\partial r_2}{\partial c_3} & \frac{\partial r_2}{\partial c_4} \\ \frac{\partial r_3}{\partial c_1} & \frac{\partial r_3}{\partial c_2} & \frac{\partial r_3}{\partial c_3} & \frac{\partial r_3}{\partial c_4} \\ \frac{\partial r_4}{\partial c_1} & \frac{\partial r_4}{\partial c_2} & \frac{\partial r_4}{\partial c_3} & \frac{\partial r_4}{\partial c_4} \end{bmatrix} = \begin{bmatrix} \overline{u}_1 & \overline{u}_1^2 & \overline{u}_1^3 & \overline{u}_1 e^{\overline{u}_1 c_4} \\ \overline{u}_2 & \overline{u}_2^2 & \overline{u}_2^3 & \overline{u}_2 e^{\overline{u}_2 c_4} \\ \overline{u}_3 & \overline{u}_3^2 & \overline{u}_3^3 & \overline{u}_3 e^{\overline{u}_3 c_4} \\ \overline{u}_4 & \overline{u}_4^2 & \overline{u}_4^3 & \overline{u}_4 e^{\overline{u}_4 c_4} \end{bmatrix}$$

which is of full rank, i.e. $\det(J^T(c)J(c)) > 0$, for each $c_4 \neq 0$, and as far as the input points $\overline{u}_1, ..., \overline{u}_4$ (freely selected by experimenter) are nonzero and different. In the experiment we set $\overline{u}_1 = -2$, $\overline{u}_2 = -1$, $\overline{u}_3 = 1$ and $\overline{u}_4 = 2$ for computing the nonparametric pointwise estimates $\widehat{w}_{1,M}, ..., \widehat{w}_{4,M}$ of the unknown $w_1 = \mu(\overline{u}_1, c^*), ..., w_4 = \mu(\overline{u}_4, c^*)$. For $\lambda_0 = 1/1024$ and $\nu = 8$, after 100 iterations, for each M ranging over $100, ..., 1000$ and $\delta = 0.1, 0.2, 0.3$, we obtained the estimation errors $\Delta_c(M) = \frac{\left\| \widehat{c}_{N_0,M} - c^* \right\|_2}{\|c^*\|_2} \cdot 100\%$ shown in Fig. 2.7. The plots illustrate rather small dependence of the estimation accuracy on the number of data M used for nonparametric estimation of interaction inputs w_n in Stage 1 for each examined intensity of noise δ when $M > 800$, and show in particular that $\Delta_c(M) < 5\%$ is guaranteed for each δ if $M > 400$ observations.

The error of the estimate (2.23) of the linear dynamics parameters θ was computed for various strategies of the choice of instrumental variables and

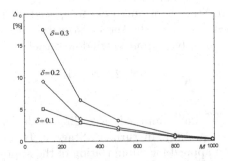

Fig. 2.7 Estimation error for Levenberg-Marquardt method

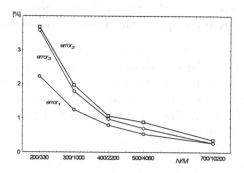

Fig. 2.8 Linear dynamics estimation error versus number of measurements

$R = 10$ independent trials for each strategy. For each instrumental variables choice, we assumed that $M = bN^{2.75}$ with the scale factor $b = 1.54 \cdot 10^{-4}$ (see Theorem 2.6) using the rule proposed in [64] (see Section VI, p. 1373) and changed the data length M over $330, ..., 10200$. This yielded the number of instruments N (see (2.24)) ranging, respectively, from 200 to 700, and the ratio N/M as shown in Fig. 2.8. First, the perfect optimal instruments Ψ_N^* (2.37) were used under working hypothesis of accessibility of signals $\{w_k\}$ and $\{v_k\}$ in the system. The computed $error_1(N) = \max_{r=1,...,R} \frac{\left\|\widehat{\theta}_N^{(r)}(\Psi_N^{*(r)}) - \theta\right\|_2}{\|\theta\|_2} \cdot 100\%$ may be clearly treated as an empirical lower bound of the estimation errors for other choices of instrumental variables (see (2.38)). Next, the strategy of nonparametric generation of instruments (2.31) has been applied using kernel estimates of w_k's (of the same form as for the static element), and the $error_2(N) = \max_{r=1,...,R} \frac{\left\|\widehat{\theta}_{N,M}^{(r)}(\widehat{\Psi}_{N,M}^{(r)}) - \theta\right\|_2}{\|\theta\|_2} \cdot 100\%$ was computed. Finally, the suboptimal instruments given by (2.39) have been applied with similar estimates of w_k's, and the impulse response cut-off level $F = 4$, yielding the $error_3(N) = \max_{r=1,...,R} \frac{\left\|\widehat{\theta}_{N,M}^{(r)}(\widehat{\Psi}_{N,M}^{*(r)}) - \theta\right\|_2}{\|\theta\|_2} \cdot 100\%$. The above three errors are shown in Fig. 2.8. According to our expectation, the best – however not realizable – is the choice of $\Psi_N = \Psi_N^*$. The "empirical" strategies $\widehat{\Psi}_{N,M}$ and $\widehat{\Psi}_{N,M}^*$ are worse, and in the sense of the relative error used in the experiment rather indistinguishable, but the same level of estimation accuracy as for the static system, i.e. not less than 5%, is achieved already for $M = 330$ input-output data and $N = 200$ instruments. Moreover, the difference between the relative estimation errors for the proposed "empirical" and optimal instruments is in general not greater than 1.5%. Hence, in our example, all strategies of instruments selection are in fact comparable, which in particular justifies the proposed strategies based on nonparametric estimation of instruments.

Nonlinearity Recovering of the Hammerstein System under Incorrect a Priori Knowledge of the Linear Dynamics

In the second experiment, we compared efficiency of our two-stage routine with the instrumental variables approach proposed by Söderström and Stoica in [131] (called further SS algorithm) in the case when a priori knowledge about the structure of the linear dynamics is inaccurate. In the Hammerstein system we assumed the linear-in-the-parameters static nonlinearity (as needed for comparison purposes)

$$\mu(u,c) = c_1 u + c_2 u^2 + c_3 u^3 \overset{c=c^*}{=} u + u^2 - u^3,$$

with the unknown parameter vector $c = c^* = (1,1,-1)^T$ and the dynamic part of the form

$$v_k = 0.5 v_{k-1} + 0.25 v_{k-1} + w_k + w_{k-1},$$

but took for the identification purposes the incorrect parametric model of the linear dynamics in the form

$$v_k = a_1 v_{k-1} + b_0 w_k.$$

The i.i.d. system input and output noise were generated as $u_k \sim U(-10, 10)$ and $\varepsilon_k \sim U(-1, 1)$, and the aim was only to estimate true parameters c^* of the nonlinearity $\mu(u, c)$. To compare efficiency of our two stage identification algorithm with the SS instrumental variables given in [131], the overall input-output parametric model of Hammerstein system needed in [131] has been computed

$$v_k^{(m)} = A_1 v_{k-1} + B_{01} u_k + B_{02} u_k^2 + B_{03} u_k^3 \triangleq \phi_k^{(ss)T} p,$$

where

$$\phi_k^{(ss)} = (v_{k-1}, u_k, u_k^2, u_k^3)^T, \qquad \text{and} \qquad p = (p_1, p_2, p_3, p_4)^T,$$

with

$$p_1 = A_1 = a_1, \quad p_2 = B_{01} = b_0 c_1, \quad p_3 = B_{02} = b_0 c_2, \quad p_4 = B_{03} = b_0 c_3,$$

and implemented in the IV estimate described in [131]:

$$\widehat{p}_N^{(ss)} = (\Psi_N^{(ss)T} \Theta_N)^{-1} \Psi_N^{(ss)T} Y_N, \tag{2.55}$$

where $\Psi_N^{(ss)} = (\psi_1^{(ss)}, \psi_2^{(ss)}, ..., \psi_N^{(ss)})^T$ and the instruments are $\psi_k^{(ss)} = (u_{k-1}, u_k, u_k^2, u_k^3)^T$. In turn, owing to the particular linear-in-the-parameters

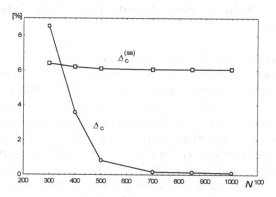

Fig. 2.9 Comparison of the two-stage parametric-nonparametric and SS algorithm

form of the nonlinearity, according to the discussion in [64] (see Remark 2 there), for performing the nonparametric stage in our two-stage procedure (Stage 1) we set $N_0 = 10 \geqslant \dim c = 3$, and arbitrarily take the input points $\overline{u}_i = -9 + 2\,(i - 1)$ forming an equidistant grid on the input domain $(-10, 10)$, and moreover guaranteeing fulfillment of the condition $rank \Phi_{N_0} = \dim c$ where $\Phi_{N_0} = (\phi_1, \phi_2, ..., \phi_{N_0})^T$, $\phi_k = (\overline{u}_k, \overline{u}_k^2, \overline{u}_k^3)^T$. This is because for the static nonlinearity as in the example in the parametric stage (Stage 2) the linear least squares can be used getting the estimate ([64])

$$\widehat{c}_{N_0, N} = \left(\Phi_{N_0}^T \Phi_{N_0}\right)^{-1} \Phi_{N_0}^T \widehat{W}_{N_0, N},$$

where $\widehat{W}_{N_0, N} = (\widehat{w}_{1,N}, \widehat{w}_{2,N}, ..., \widehat{w}_{N_0, N})^T$ with the nonparametric kernel estimates $\widehat{w}_{k,N}$. For comparison purposes, in the kernel estimates of w_n's (see (2.50), (2.54)) we assumed $M = N$ (with N the same as for (2.55)) taking due to [49] $h_{opt}(N) = 4.1 N^{-0.2}$. Next we compared appropriate relative estimation errors of the nonlinearity parameters produced by both methods

$$\Delta_c = \frac{\|\widehat{c}_{N_0, N} - c^*\|_2}{\|c^*\|_2} \cdot 100\%, \quad \text{and} \quad \Delta_c^{(ss)} = \frac{\left\|\widehat{p'}_N^{(ss)} - c^*\right\|_2}{\|c^*\|_2} \cdot 100\%,$$

where $\widehat{p'}_N^{(ss)} = (\frac{\widehat{p}_{2,N}^{(ss)}}{b_0}, \frac{\widehat{p}_{3,N}^{(ss)}}{b_0}, \frac{\widehat{p}_{4,N}^{(ss)}}{b_0})^T$ for growing data length N. The results are presented in Fig. 2.9. We see that our combined parametric-nonparametric approach is here truly robust to the incorrect knowledge of the parametric description of the linear subsystem, in contrast to the SS algorithm in [131], where the essential systematic estimation error (bias) appears.

2.4 Hammerstein System Excited by Correlated Process

In this section we assume that the input is not i.i.d. sequence, but the system characteristic is of polynomial form, with known order (see [98]). The goal is to identify nonlinearity parameters when the system is excited and disturbed by correlated random processes. The problem is semi-parametric in the sense that the nonlinear static characteristic is recovered without prior knowledge about the linear dynamic block, i.e., when its order is unknown. The method is based on the instrumental variables technique, with the instruments generated by input inverse filtering. It is proved that, in contrary to the least squares based approach, the proposed algorithm leads to an asymptotically unbiased, strongly consistent estimate. Constructive procedures of instrumental variables generation are given for some popular cases.

2.4.1 Introduction

The class of approximating structures admitted here includes Hammerstein systems and their parallel connections. The significant difference of our approach, in comparison with the algorithms known in the literature, lies in the fact that

- the input process of the nonlinear block can be *colored* with an arbitrary correlation structure, and moreover
- the linear dynamic block(s) in a modelling system is stable, but its order remains *unknown.*

The main advantage of the approach is that we obtain a consistent estimate of the nonlinear characteristic of the system, when the prior knowledge is *semi-parametric* (partial), i.e., with nonparametric treatment of the linear dynamic block (see e.g. [46] and [71]), and in a *correlated input* case. Since, in such a case there is no simple relation between the identified characteristic and the regression function, the least squares algorithm fails. It is a well known fact that the least squares (*LS*) approach to system identification can give asymptotically biased estimates ([62], [65], [154]). In many algorithms, considered in the literature, the reason of the non-vanishing bias is the cross-correlation between the noise and the elements of 'generalized' input vectors, which include previous outputs (for details, we refer the reader to Appendix A.23, where a simple example from linear system theory is given). Its various modifications (see [33], [127]–[129], [150], [157], [156], [154]) deal with reducing the bias caused by correlation of the noise, however, the semi-parametric problem with correlated input, in spite of occurring in practice, has been treated very cursorily in the literature.

Since the correlation of the input may also be caused by the structural feedback in a complex system (the element excitation by the output of the preceding subsystem (see [62]) and Section 6), the above assumptions seem to be of high practical importance.

We consider a class of block-oriented systems including various structures built of one static nonlinear element with the characteristic $\mu()$, and one or more linear dynamic blocks ([141]). Only the input x_k and the noise-corrupted output y_k of the whole complex system are accessible, and the internal signals are hidden. Our aim is to recover the function $\mu()$ (or its scaled version) from the observations $\{(x_k, y_k)\}$. Similarly as in [65], we assume the parametric knowledge of the identified static characteristic, but, in contrast to other approaches (see e.g. [101], [113], [131], [145]), the orders of the linear dynamic blocks are assumed to be unknown. The nonparametric algorithms [50], which are dedicated for the problems with the lack of prior knowledge about the linear blocks, are based on the observation that for the considered class of systems (defined in Section 2.4.2), and for white input, the regression function

$$R(x) \triangleq E\{y_k | x_k = x\}$$

equals to the system characteristic $\mu(x)$ (up to a constant offset and scaling factor). Unfortunately, regression-based nonparametric estimates (see [50]) do not converge to the system characteristic, when the input process is correlated (with the exception of the trivial case of a static system), which is serious limitation in practice. In Tables 3.1 and 3.2 we show how the single dimensional regression $R(x)$ and the multiple regression

$$R(x, v) \triangleq E\{y_k | x_k = u, x_{k-1} = v\}$$

depend on the system characteristic $\mu(x)$ for a simple static system and for the Hammerstein system (see Fig. 2.11), with FIR(2) linear block, disturbed by zero-mean random output noise z_k. The table illustrates the fact that the correlated input process is the reason of the specific problem, when the true system is not static and the prior knowledge of the linear dynamic component is not accessible (i.e., $e(x) \neq$ const, if the order of assumed model of dynamic component is too small).

Table 2.1 Regression function vs. the nonlinearity for static system

Static system $y_k = \mu(x_k) + z_k$	
i.i.d. input x_k	colored input x_k
$R(x) = \mu(x)$	$R(x) = \mu(x)$
$R(x, v) = \mu(x)$	$R(x, v) = \mu(x)$

In the traditional (purely parametric) approaches, the input need not to be an i.i.d. process for the LS estimate to be consistent, thanks to restrictive assumptions imposed on the linear dynamic block. In the problem considered here, the linear component(s) is unknown and treated as a black-box.

Table 2.2 Regression function vs. the nonlinearity for FIR(2) Hammerstein system

Hammerstein system $y_k = \gamma_0\mu(x_k) + \gamma_1\mu(x_{k-1}) + z_k$	
i.i.d. input x_k	colored input x_k
$R(x) = \gamma_0\mu(x) + e$	$R(x) = \gamma_0\mu(x) + e(x)$
$e = \gamma_1 E\mu(x_{k-1}) = \text{const}$	$e(x) = \gamma_1 E\{\mu(x_{k-1})\|x_k = x\} \neq \text{const}$
$R(x,v) = \gamma_0\mu(x) + \gamma_1\mu(v)$	$R(x,v) = \gamma_0\mu(x) + \gamma_1\mu(v)$

This generalization makes the problem of nonlinearity estimation much more difficult and leads to new and original application of instrumental variables technique.

In Sections 2.4.2-2.4.4 the class of considered nonlinear dynamic systems is defined, and the identification problem is formulated in detail. We focus there on the recovering of the nonlinear characteristic of a static element. Next, in Section 2.4.5, the least squares estimate is analyzed and the reason of its bias, in the correlated input case with semi-parametric knowledge, is exhibited. In order to reduce that bias, the generalized instrumental variables estimate is proposed in Section 2.4.6.

We generalize the idea of instrumental variables approach in the sense that the proposed method allows to eliminate the bias of the nonlinearity estimate caused by correlation of both the noise and input processes of a nonlinear system. The originality of the approach is that the estimation of the input correlation structure and input inverse filtering (e.g., by the blind decon-volution) supports the generation of instrumental variables, which guarantee consistency of the nonlinearity estimate, without necessity of parametrization of the linear dynamic component. The consistency conditions are formulated and proved in Section 2.4.7 and, finally, constructive procedures of the gen-eration of instruments are given in Section 2.4.8.

2.4.2 Assumptions

Consider a class of nonlinear block-oriented dynamic systems, described by the following equation (cf. [107])

$$y_k = F(x_k) + \xi_k(x_{k-1}, x_{k-2}, ...) + z_k, \tag{2.56}$$

where $F()$ is an unknown nonlinear static characteristic,

$$\xi_k = \xi_k(x_{k-1}, x_{k-2}, ...)$$

represents the dynamic properties and is treated as a 'system disturbance', and $\{z_k\}$ is a measurement noise. Moreover, let us assume that:

Assumption 2.16. *The input $\{x_k\}$ is a stationary, correlated process, which can be interpreted as the output of asymptotically stable $(\Sigma_{i=0}^{\infty} |g_i| < \infty)$ linear filter*

$$x_k = \sum_{i=0}^{\infty} g_i u_{k-i} \qquad (2.57)$$

with the unknown impulse response $\{g_i\}$. The random process $\{u_k\}$ (also unknown) is assumed to be bounded $(|u_k| < u_{\max} < \infty)$ and i.i.d. For clarity of the presentation and without any loss of generality we assume that $g_0 = 1$.

Assumption 2.17. *The nonlinear function $F()$ is the polynomial*

$$F(x) = \sum_{w=1}^{W} p_w x^{w-1} \qquad (2.58)$$

of known order $W - 1$, and p_w $(w = 1, ..., W)$ are unknown parameters.

Assumption 2.18. *The 'system noise' $\{\xi_k\}_{k \in Z}$ can be written in the form*

$$\xi_k = \sum_{i=1}^{\infty} \gamma_i \zeta(x_{k-i}), \qquad (2.59)$$

where $\zeta(x)$ is unknown, bounded function

$$|\zeta(x)| \leq M_\zeta < \infty, \qquad (2.60)$$

such that $E\zeta(x_1) = 0$. The sequence $\{\gamma_i\}$ depends on the unknown impulse response of the linear blocks in the complex system.

Assumption 2.19. *The measurement noise $\{z_k\}_{k \in Z}$ is the stationary, correlated, zero-mean random process, which can be interpreted as the output of an unknown linear filter with impulse response $\{\omega_i\}_{i=0}^{\infty}$, where $\Sigma_{i=0}^{\infty} |\omega_i| < \infty$, excited by the zero-mean $(E\varepsilon_k = 0)$, bounded white noise $\{\varepsilon_k\}$ $(|\varepsilon_k| < \varepsilon_{MAX})$.*

Assumption 2.20. *Processes $\{x_k\}$ and $\{\varepsilon_k\}$ are mutually independent.*

Assumption 2.21. *Only $\{x_k\}$ and $\{y_k\}$ can be measured.*

2.4.3 Examples

It exists a number of block-oriented nonlinear systems which fall into the above description (i.e., static element (Fig. 2.10), Hammerstein system (Fig. 2.11), parallel system (Fig. 2.12), parallel-series system (see Fig. 2.13 and [69], [107])). For these systems the nonlinearity $F(x)$ in equation (2.56) is in general a scaled and shifted version of the true system nonlinearity $\mu(x)$, i.e.,

$$F(x) = c\mu(x) + l, \qquad (2.61)$$

where c and l are constants, impossible to identify, dependent on the particular system structure. For example in Hammerstein system (Fig. 2.11) we get

$$y_k = \sum_{i=0}^{\infty} \gamma_i \mu\left(x_{k-i}\right) + z_k = \gamma_0 \mu\left(x_k\right) + E\mu\left(x_1\right) \sum_{i=1}^{\infty} \gamma_i +$$

$$+ \sum_{i=1}^{\infty} \gamma_i \left[\mu\left(x_{k-i}\right) - E\mu\left(x_1\right)\right] + z_k,$$

which leads to

$$c = \gamma_0, \qquad l = E\mu\left(x_1\right) \sum_{i=1}^{\infty} \gamma_i, \qquad \zeta(x) = \mu(x) - E\mu\left(x_1\right). \qquad (2.62)$$

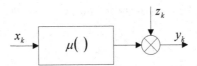

Fig. 2.10 Static system

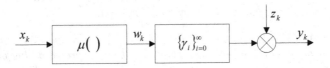

Fig. 2.11 Hammerstein system

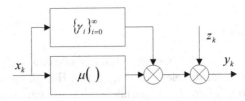

Fig. 2.12 Parallel system

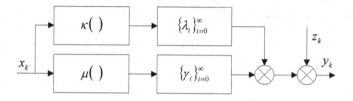

Fig. 2.13 Parallel-series system

Remark 2.10. *In the considered class of systems, the constant c cannot be identified independently of the identification method. It is consequence of inaccessibility of internal signals for a direct measurement. For example, the Hammerstein systems with the characteristics $c\mu(x)$ and the linear blocks $\{\frac{1}{c}\gamma_i\}$ are, for various c, indistinguishable from the input-output point of view. Owing to above, we can assume without loss of generality that $c = 1$.*

Remark 2.11. *The unknown offset l is a consequence of the lack of prior knowledge of the system structure. It is a 'cost paid' for the universality of the approach, i.e., the whole block-oriented system can be treated in the same manner as the static one. However, if the value of the static characteristic is given in at least one point x_0 (e.g., we know that $\mu(0) = 0$), then for $c = 1$ we obtain*

$$\mu(x) = F(x) - F(x_0) - \mu(x_0),$$

which allows to avoid the offset l.

Remark 2.12. *From (2.61) we conclude that polynomial form of $\mu()$ implies fulfilment of Assumption 2.17 (i.e. polynomial form of $F(x)$, of the same order $W - 1$).*

2.4.4 Statement of the Problem

Taking (2.56) and (2.58) into consideration, and introducing the generalized input vector

$$\overline{x}_k = \left(1, x_k, x_k^2, ..., x_k^{W-1}\right)^T, \tag{2.63}$$

the parameter vector

$$p = (p_1, ..., p_W)^T, \tag{2.64}$$

and the total disturbance

$$d_k = z_k + \xi_k(x_{k-1}, x_{k-2}, ...), \tag{2.65}$$

we obtain that

$$y_k = \overline{x}_k^T p + d_k,$$

and appropriate matrix-vector compact description

$$Y_N = X_N p + D_N, \tag{2.66}$$

where

$$Y_N = (y_1, ..., y_N)^T, \tag{2.67}$$
$$X_N = (\bar{x}_1, ..., \bar{x}_N)^T, \tag{2.68}$$
$$D_N = (d_1, ..., d_N)^T. \tag{2.69}$$

According to Assumption 2.18, it holds that $E\xi_k(x_{k-1}, x_{k-2}, ...) = 0$, and on the basis of Assumption 2.19 we obtain that $Ez_k = 0$, which yields to

$$Ed_k = 0, \quad \text{for } k = 1, 2, ..., N. \tag{2.70}$$

The aim of the identification is to recover parameters p_w $(w = 1, ..., W)$ of $F(x)$ (see (2.58) and (2.64)), using the set of input-output measurements $\{(x_k, y_k)\}_{k=1}^N$ of the whole complex system, obtained in the experiment.

2.4.5 Semi-parametric Least Squares in Case of Correlated Input

In the considered semi-parametric problem, the least squares based estimate has the form

$$\widehat{p}_N^{(LS)} = \left(\frac{1}{N}X_N^T X_N\right)^{-1}\frac{1}{N}X_N^T Y_N. \tag{2.71}$$

Inserting (2.66) into (2.71) we obtain that $\widehat{p}_N^{(LS)} = p + \Delta_N^{(LS)}$, where

$$\Delta_N^{(LS)} = \left(\frac{1}{N}X_N^T X_N\right)^{-1}\frac{1}{N}X_N^T D_N \tag{2.72}$$

is the estimation error which, as it is shown below, does not tend to zero as $N \to \infty$. By Assumptions 2.16, 2.18 and 2.19, the disturbance d_k and elements of $\bar{x}_k$ are bounded, i.e., $|d_k| < d_{\max} < \infty$, $\left|x_k^{w-1}\right| < x_{\max}^{w-1} < \infty$, where $d_{\max} = \sum_{i=0}^\infty |\omega_i| u_{\max} + \sum_{i=1}^\infty |\gamma_i| M_\varsigma$ and $x_{\max} = \sum_{i=0}^\infty |g_i| u_{\max}$. Therefore

$$|Ed_k| < \infty, \qquad \text{var} d_k < \infty, \tag{2.73}$$
$$\left|Ex_k^{w-1}\right| < \infty, \qquad \text{var} x_k^{w-1} < \infty,$$

for each $w = 1, ..., W$. In particular, the 4th order moments are finite

$$\left|Ed_k^4\right| < \infty, \qquad \left|Ex_k^{4(w-1)}\right| < \infty \tag{2.74}$$

which will be essential for further conclusions. Under Assumption 2.16, x_k and $x_{k+\tau}$ become asymptotically independent (as $|\tau| \to \infty$), so the autocorrelation of $\{x_k^{w-1}\}$ for each $w = 1, ..., W$ has the property

$$r_{x_k^{w-1}}(\tau) = Ex_k^{w-1}x_{k-\tau}^{w-1} - Ex_k^{w-1}Ex_{k-\tau}^{w-1} \to 0, \qquad \text{as } |\tau| \to \infty. \qquad (2.75)$$

Directly from (2.65) and Assumptions 2.16, 2.18 and 2.19 we obtain that

$$r_d(\tau) \to 0, \qquad \text{as } |\tau| \to \infty. \qquad (2.76)$$

From (2.73), (2.75) and Lemma B.11 in Appendix B.8 we conclude that

$$\frac{1}{N}X_N^T X_N = \frac{1}{N}\sum_{k=1}^{N} \bar{x}_k \bar{x}_k^T \to E\bar{x}_1\bar{x}_1^T \qquad (2.77)$$

with probability 1 as $N \to \infty$ (see [105]). Similarly, by virtue of (2.74), (2.75), (2.76) and Lemma B.11

$$\frac{1}{N}X_N^T D_N = \frac{1}{N}\sum_{k=1}^{N} \bar{x}_k d_k \to \begin{bmatrix} Ed_1 \\ Ex_1 d_1 \\ Ex_1^2 d_1 \\ \vdots \\ Ex_1^{W-1}d_1 \end{bmatrix} = E\bar{x}_1 d_1 \qquad (2.78)$$

with probability 1, as $N \to \infty$. Taking advantage of the property that ARMA processes are SPE of any finite orders (see Definition B.6 in Appendix B.7), the limit matrix in (2.77) is nonsingular, and LS estimate (2.71) is well defined. However, since $\{x_k\}$ is colored (see Assumption 2.16), x_k and d_k are in general statistically dependent, and hence $Ex_1^{w-1}d_k \neq 0$, although $Ed_k = 0$ (see (2.70)). Consequently, LS method provides the estimate, which is asymptotically biased, i.e.,

$$\hat{p}_\infty^{(LS)} = p + \left(E\bar{x}_1\bar{x}_1^T\right)^{-1}E\bar{x}_1 d_1,$$

because of the correlation between x_k^{w-1}'s (occurring in $\bar{x}_k$) and $d_k = z_k + \xi_k(x_{k-1}, x_{k-2}, ...)$.

2.4.6 Instrumental Variables Method

In order to remove the bias, assume that there exists $\Psi_N = (\psi_1, ..., \psi_N)^T$, where $\psi_k \triangleq (\psi_{k,1}, ..., \psi_{k,W})^T$ (instrumental variables matrix), so the following conditions are fulfilled:

(C1) $\frac{1}{N}\Psi_N^T X_N \to E\psi_1\bar{x}_1^T$ with probability 1, and the limit matrix is nonsingular, i.e.,

$$\det\{E\psi_1\bar{x}_1^T\} \neq 0,$$

(C2) $\frac{1}{N}\Psi_N^T D_N \to E\psi_1 d_1$ with probability 1, and $E\psi_1 d_1 = 0$,

and consider the IV-estimate of the form

$$\hat{p}_N^{(IV)} = \left(\frac{1}{N}\Psi_N^T X_N\right)^{-1} \frac{1}{N}\Psi_N^T Y_N. \qquad (2.79)$$

Inserting (2.66) into (2.79) we obtain the formula for the appropriate estimation error

$$\Delta_N^{(IV)} = \hat{p}_N^{(IV)} - p = \left(\frac{1}{N}\Psi_N^T X_N\right)^{-1} \frac{1}{N}\Psi_N^T D_N, \qquad (2.80)$$

and further by Slutzky theorem it holds that

$$\mathrm{plim}_{N\to\infty}\Delta_N^{(IV)} = \mathrm{plim}_{N\to\infty}\left(\frac{1}{N}\Psi_N^T X_N\right)^{-1}\mathrm{plim}_{N\to\infty}\left(\frac{1}{N}\Psi_N^T D_N\right).$$

From the continuity of the matrix inversion operator we conclude that

$$\mathrm{plim}_{N\to\infty}\left(\frac{1}{N}\Psi_N^T X_N\right)^{-1} = \left(\mathrm{plim}_{N\to\infty}\left(\frac{1}{N}\Psi_N^T X_N\right)\right)^{-1},$$

and on the basis of (C1) the term is well defined. Thus from (C2) we have $\mathrm{plim}_{N\to\infty}\Delta_N^{(IV)} = 0$, and by boundedness of all processes the strong convergence also holds. Consequently, if (C1) and (C2) are fulfilled, then the estimate given by (2.79) is strongly consistent, i.e.

$$\hat{p}_N^{(IV)} \to p \text{ with probability 1, as } N \to \infty, \qquad (2.81)$$

independently of the input correlation structure.

The fundamental question is the manner of Ψ_N-generation in order to assure fulfilment of (C1) and (C2). It means that the instruments should be correlated with inputs x_k (see (C1)) and simultaneously not correlated with generalized disturbances d_k (see (C2)), that are dependent on the previous inputs $x_{k-1}, x_{k-2}, \ldots$. Several methods of generation of Ψ_N, appropriate for some important cases are given below.

2.4.7 Generation of Instruments

In this Section we develop the issue of generation of instrumental variables. First, we assume the accessibility of the signal $\{u_k\}$ to show that it can be used for construction of the matrix Ψ_N, which guarantees (C1) and (C2) and next, we consider the problem formulated above, when only $\{x_k\}$ is accessible, and $\{u_k\}$ must be generated (see Assumption 2.21). In general, the presented idea is based on blind deconvolution (see e.g. [17], [31]) of the input process $\{x_k\}$, which leads to the two-stage routine.

Known $\{u_k\}$

Let us consider the structure in Fig. 2.14, where the signal $\{u_k\}$ is accessible for a direct measurement, but the impulse response $\{g_i\}_{i=0}^{\infty}$ of the linear filter which forms $\{x_k\}$ is still unknown. Such a problem resembles the classical Wiener-Hammerstein (sandwich) identification task. Nevertheless, the structure of the identified block-oriented system, and the orders of the linear dynamics included therein, are unknown. The following theorem holds.

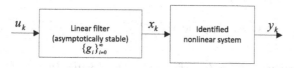

Fig. 2.14 Cascade structure

Theorem 2.9. *The instrumental variable matrix Ψ_N generated according to the rule*

$$\Psi_N = (\psi_1, \psi_2, ..., \psi_N)^T, \text{ where } \psi_k = \overline{u}_k = \left(1, u_k, u_k^2, ..., u_k^{W-1}\right)^T, \quad (2.82)$$

fulfils (C1) *and* (C2).

Proof. For the proof see Appendix A.11. ∎

Unknown $\{u_k\}$

On the basis of Theorem 2.9 the estimate $\widehat{p}_N^{(IV)} = \left(\frac{1}{N}\Psi_N^T X_N\right)^{-1} \frac{1}{N}\Psi_N^T Y_N$ with $\psi_k = \overline{u}_k$ is strongly consistent, i.e., converges with probability 1 to the parameter vector p of the identified function $F(x)$. Nevertheless, since the process $\{u_k\}$ cannot be measured, its estimation is needed.

Hence, we propose the preliminary deconvolution of $\{x_k\}$ to provide estimates $\widehat{u}_k$ of u_k. Let moreover $\widehat{u}_k$ be computed using the sequence $\{x_k\}_{k=1}^M$ of M measurements, which in general is greater than N. The following theorem shows the relation between N and M, which guarantees the consistency of the two-stage estimate.

Theorem 2.10. *If the estimation error of $\widehat{u}_{k,M}$ behaves like*

$$|\widehat{u}_{k,M} - u_k| = O(M^{-\tau}) \text{ with probability 1 as } M \to \infty, \quad (2.83)$$

then for the instruments

$$\Psi_{N,M} = (\psi_{1,M}, \psi_{2,M}, ..., \psi_{N,M})^T, \quad (2.84)$$

where $\psi_{k,M} = \widehat{\overline{u}}_{k,M} = \left(1, \widehat{u}_{k,M}, \widehat{u}_{k,M}^2, ..., \widehat{u}_{k,M}^{W-1}\right)^T$, *the two-stage estimate*

$$\widehat{p}_{N,M}^{(IV)} = \left(\frac{1}{N} \Psi_{N,M}^T X_N \right)^{-1} \frac{1}{N} \Psi_{N,M}^T Y_N \tag{2.85}$$

is strongly consistent, i.e.,

$$\widehat{p}_{N,M}^{(IV)} \to p \ \textit{with probability} \ 1 \ \textit{as} \ N, M \to \infty,$$

provided that $NM^{-\tau} \to 0$.

Proof. For the proof see Appendix A.12.   ∎

2.4.8 *Various Kinds of Prior Knowledge*

We present several strategies of deconvolution which can be applied for typical cases in practice. The choice of appropriate method depends on the prior knowledge of the excitation $\{x_k\}$.

ARX Input

Let the input process $\{x_k\}$ is assumed to be of the ARX(s) form, i.e.,

$$x_k = a_1 x_{k-1} + a_2 x_{k-2} + \ldots + a_s x_{k-s} + u_k, \tag{2.86}$$

where $a_1, a_2, \ldots, a_s$ are unknown parameters and the process $\{u_k\}$ cannot be measured. For clarity of the exposition, and without any loss of generality, let $Eu_k = 0$ (for the case of $Eu_k \neq 0$ see Section IV in [92]). Introducing the regressor $\phi_k = (x_{k-1}, x_{k-2}, \ldots, x_{k-s})^T$, and the vector of unknown parameters

$$\theta = (a_1, a_2, \ldots, a_s)^T, \tag{2.87}$$

we get $x_k = \phi_k^T \theta + u_k$, $k = 1, 2, \ldots, M$, or in the compact, matrix-vector version $X_M = \Phi_M \theta + U_M$, where $X_M = (x_1, x_2, \ldots, x_M)^T$, $\Phi_M = (\phi_1, \phi_2, \ldots, \phi_M)^T$ and $U_M = (u_1, u_2, \ldots, u_M)^T$. The unknown part $U_N = (u_1, u_2, \ldots, u_N)^T$, needed for generation of instruments, can be estimated with the help of the least squares method. The results are then plugged into (2.84), which leads to the following scheme:

Stage 1. Compute

$$\widehat{U}_{N,M} = X_N - \Phi_N \widehat{\theta}_M, \text{ where } \widehat{\theta}_M = \left(\Phi_M^T \Phi_M \right)^{-1} \Phi_M^T X_M. \tag{2.88}$$

Stage 2. Using the estimates $\widehat{U}_{N,M} = (\widehat{u}_{1,M}, \widehat{u}_{2,M}, \ldots, \widehat{u}_{N,M})^T$ obtained in *Stage 1*, generate the instrumental variable matrix $\Psi_{N,M}$ according to (2.84), and compute (2.85).

Remark 2.13. *For the estimate (2.88) we have standard parametric convergence rate* $|\widehat{u}_{k,M} - u_k| = O(M^{-1/2})$, *i.e.,* $\tau = \frac{1}{2}$ *(see 2.83), and by Theorem 4 in [64], for* $M = const \cdot N^3$ *the convergence rate of the system parameter estimates is* $\left\|\widehat{p}_{N,M}^{(IV)} - p\right\| = O(N^{-1/2})$ *with probability 1.*

Gauss-Markov Approach

Assume that the autocorrelation function $r_x(\tau) = E\left\{(x_k - Ex_1)((x_{k+\tau} - Ex_1))\right\}$ of the input $\{x_k\}$ is a priori known, and use the covariance matrix

$$\Sigma_M = \begin{bmatrix} r_x(0) & r_x(1) \ . & \ . & r_x(M-1) \\ r_x(1) & r_x(0) \ r_x(1) \ . & \ . & \\ . & r_x(1) \ r_x(0) \ r_x(1) \ . & & \\ . & & r_x(1) \ r_x(0) \ r_x(1) & \\ r_x(M-1) \ . & \ . & r_x(1) \ r_x(0) \end{bmatrix}.$$

According to Theorem B.2 in Appendix B.2 one can find the root matrix P_M, such that $P_M P_M^T = \Sigma_M$. For

$$\widetilde{U}_{N,M} = P_M^{-1} X_N$$

it can easily be shown that asymptotically, as $M \to \infty$, it holds that

$$E\widetilde{U}_{N,M}\widetilde{U}_{N,M}^T = I \text{ with probability 1.}$$

Therefore, the elements of $\widetilde{U}_{N,M}$ may be used analogously as the sequence $\{u_k\}$ in (2.82).

Deconvolution by Cummulant Optimization

In a general case, if the filter (2.57), which forms the input is invertible (i.e. all zeros of $G(z) = \mathcal{Z}\{g_i\}$ lies in the unit circle), then its inversion is stable and can be approximated, with an arbitrarily small error, by the model

$$v_k(b) = b_0 x_k + b_1 x_{k-1} + ... + b_s x_{k-s}, \tag{2.89}$$

with the finite impulse response $b = (b_0, b_1, ..., b_s)^T$. Since for large enough s this error can be neglected, we assume that there exists $b^* \in R^s$, such that $v_k(b^*)$ is a white noise, and maximize absolute of any-order normalized cummulant of $v_k(b)$ with respect to b (see e.g. [17]). Since for all nontrivial random variable x_k it holds that $\mathrm{var} v_k(b) \neq 0$, the most natural idea is analysis of kurtosis, i.e. $\kappa_v(b) = \frac{\mu_4(b)}{\sigma^4(b)} - 3$, where $\mu_4(b)$ and $\sigma(b)$ are respectively the 4th central moment and the standard deviation of $v_k(b)$. It leads to the following generation procedure

$$\psi_{k,M} = \left(1, v_k(\widehat{b}), v_k^2(\widehat{b}), ..., v_k^{W-1}(\widehat{b})\right)^T,$$

where

$$\widehat{b} = \arg\max_b |\widehat{\kappa}_v(b)|, \qquad (2.90)$$

and $\widehat{\kappa}_v(b)$ is the estimate of kurtosis measure, specified by

$$\widehat{\kappa}_v(b) = \frac{\frac{1}{M}\sum_{k=1}^{M}(v_k(b) - \overline{v}(b))^4}{\left(\frac{1}{M}\sum_{k=1}^{M}(v_k(b) - \overline{v}(b))^2\right)^2} - 3,$$

with $\overline{v}(b) = \frac{1}{M}\sum_{k=1}^{M} v_k(b)$.

2.4.9 Computer Experiment

Simulation

The above algorithms were implemented and compared on the simple example of Hammerstein system to illustrate performance of the semi-parametric least squares and instrumental variables algorithms. The polynomial static characteristic

$$w_k = \mu(x_k) = x_k^3, \qquad (2.91)$$

followed by the IIR linear dynamic block

$$\overline{y}_k = 0.5\overline{y}_{k-1} + 0.5w_k, \qquad (2.92)$$

were used in the simulation. Infinite memory input excitation $\{x_k\}$ was generated, according to $x_k = 0.5x_{k-1} + 0.5u_k$, with uniformly distributed innovation $u_k \sim U(-1,1)$. The output $\overline{y}$ was contaminated by the random noise $\varepsilon_k \sim U(-0.1, 0.1)$, i.e., we obtained the pairs $\{(x_k, y_k)\}$, where $y_k = \overline{y}_k + \varepsilon_k$.

Identification

In the identification experiment we only assumed that:

- the unknown characteristic is polynomial of order $W - 1 \leqslant 4$ (see (2.58)),
- the structure of the identified block-oriented system is unknown (see (2.56), Assumption 2.18 and Section 2.4.3),
- the linear dynamic block(s) included in the system are stable.

Owing to (2.62), Remark 2.10 and Remark 2.11, with nonparametric treatment of the linear block(s) (no prior knowledge of its difference equation (2.92)) the true characteristic $\mu(x_k)$ can be recovered only up to unknown scale factor $c = \gamma_0 = 0.5$. Since the probability density function of x_k is even, and the characteristic $\mu()$ is odd, we obtain that $E\mu(x_k) = 0$ and $l = 0$ (see (2.61) and (2.62)), i.e.,

$$F(x_k) = 0.5x_k^3 = \overline{x}_k^T p,$$

where the parameter vector of the estimated function $F(x)$ is

$$p = (0, 0, 0, 0.5)^T$$

(see (2.58), (2.63) and (2.64)). The experiment was carried out on $N = 200, 300, ..., 5000$ simulated pairs $\{(x_k, y_k)\}_{k=1}^N$ with $M = \lfloor 0.0002 \cdot N^2 \rfloor$, and repeated $R = 100$ times for each N. Estimation error was computed according to the following rule

$$ERR(N) = \frac{1}{R} \sum_{r=1}^R \left\| \widehat{p}_N^{(r)} - p \right\|_2,$$

where $\widehat{p}_N^{(r)}$ denotes the realization of the estimate obtained in rth re-run. In Fig. 2.15 the error $ERR(N)$ of the proposed method (2.79), with various strategies of generation of instrumental variables, is compared with the error of the least squares estimate (2.71), when the number of measurements grows large. In particular, we compared the following special cases of instrumental variables generation with various prior knowledge of the process $\{x_k\}$: (a) accessibility of $\{u_k\}$ (full knowledge, see (2.82)), (b) known order of the process $\{x_k\}$, (c) known autocorrelation function of $\{x_k\}$, and (d) ψ-generation by cummulant optimization. In (2.86) and (2.89) we assumed the model of $\{x_k\}$ with the order $s = 5$ and in (2.90) we applied heuristic optimization algorithm supported by simulated annealing.

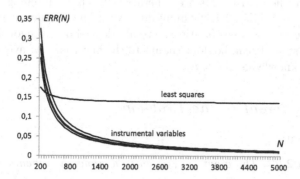

Fig. 2.15 Estimation error of the semi-parametric least squares and instrumental variables methods

2.4.10 Summary

The scope of applicability of the instrumental variables technique, which allows to eliminate the bias, is extended for a broad class of nonlinear dynamic systems excited by the colored random input, when the knowledge of dynamic blocks is poor. Several constructive strategies of generation of instrumental variables, which guarantee consistency of nonlinearity parameter estimates, are given. In general, they are based on the algorithms of blind deconvolution of the input process, which are strongly elaborated in the system identification literature. The idea presented in this section is also applicable for nonlinearity recovering in Wiener-Hammerstein sandwich systems, when the nonlinearity input is accessible for a direct measurement. The obtained results show that in real applications, i.e., when the prior knowledge of the nonlinear block-oriented system is uncertain and excitations are not white noises, the least squares method can be significantly improved by a simple generalization.

2.5 Cascade Systems

2.5.1 Introduction

This point constitutes generalization and extension of Section 2.4, where the static characteristic of Hammerstein system excited by colored linear process is identified. Here we consider a cascade connection of two Hammerstein systems, shown in Fig. 2.16, and assume that the interconnecting signal x_k can be measured. Similarly as in Section 2.4 we apply the instrumental variable method, where the instruments are generated by input whitening. However, in the contrary to Section 2.4, the input deconvolution is made with the use of nonparametric regression estimation methods (kernel or orthogonal) applied for the first system. It can bo done thanks to the fact that we have additional measurement knowledge about u_k.

2.5.2 Statement of the Problem

We assume that:

Assumption 2.22. *The input $\{u_k\}$ is an i.i.d. random sequence with finite variance.*

Assumption 2.23. *The nonlinear functions $\eta(u)$ and $\mu(x)$ are polynomials*

$$\eta(u) = \sum_{w=1}^{W} q_w u^{w-1}, \qquad \mu(x) = \sum_{w=1}^{W} p_w x^{w-1}, \qquad (2.93)$$

of known order $W - 1$, where q_w, and p_w $(w = 1, ..., W)$ are unknown parameters.

Assumption 2.24. *The linear dynamic blocks are asymptotically stable, i.e.,*

$$\sum_{i=0}^{\infty} |\lambda_i| < \infty, \qquad \sum_{i=0}^{\infty} |\gamma_i| < \infty,$$

but the impulse responses $\{\lambda_i\}_{i=0}^{\infty}$ and $\{\gamma_i\}_{i=0}^{\infty}$ are unknown.

Assumption 2.25. *The random noise processes $\{\varepsilon_k\}$ and $\{z_k\}$ are mutually independent, stationary, zero-mean, and ergodic. Moreover, both $\{\varepsilon_k\}$ and $\{z_k\}$ are independent of the external input $\{u_k\}$.*

Assumption 2.26. *Only $\{u_k\}$, $\{x_k\}$ and $\{y_k\}$ can be observed.*

Assumption 2.27. *For convenience, and without loss of generality we also assume that $\lambda_0 = 1$ and $\gamma_0 = 1$ (see [64] for discussion).*

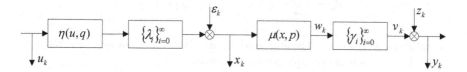

Fig. 2.16 The cascade connection of two Hammerstein systems

Remark 2.14. *The interaction signal x_k is a stationary correlated linear process, which can be interpreted as the output of asymptotically stable $(\Sigma_{i=0}^{\infty} |\lambda_i| < \infty)$ linear filter excited by the white noise $\eta(u_k)$, i.e.,*

$$x_k = \sum_{i=0}^{\infty} \lambda_i \eta(u_{k-i}) + \varepsilon_k. \tag{2.94}$$

2.5.3 Least Squares

Since the first Hammerstein system is excited by the white noise, the problem of identification of $\eta(u)$ is trivial, and standard nonparametric regression estimation methods (see e.g. [50]) can be successfully applied for the pairs $\{(u_k, x_k)\}_{k=1}^{M}$. We focus on recovering of the nonlinear characteristic $\mu(u)$ of the second system from the correlated data $\{(x_k, y_k)\}_{k=1}^{M}$. Introducing the regressors

$$\varphi_k = \left(1, u_k, u_k^2, ..., u_k^{W-1}\right)_k^T \tag{2.95}$$
$$\phi_k = \left(1, x_k, x_k^2, ..., x_k^{W-1}\right)^T,$$

and the parameter vectors

$$q = (q_1 + m, q_2, ..., q_W)^T, \tag{2.96}$$
$$p = (p_1 + l, p_2, ..., p_W)^T$$

we obtain that

$$x_k = \varphi_k^T q + e_k, \text{ and } y_k = \phi_k^T p + d_k,$$

where

$$m = E\eta(u_1) \sum_{i=1}^{\infty} \lambda_i = \text{const, and } l = E\mu(x_1) \sum_{i=1}^{\infty} \gamma_i = \text{const} \tag{2.97}$$

are unidentifiable offsets, and

$$e_k = \sum_{i=1}^{\infty} \lambda_i [\eta(u_{k-i}) - E\eta(u_1)] + \varepsilon_k, \tag{2.98}$$

$$d_k = \sum_{i=1}^{\infty} \gamma_i [\mu(x_{k-i}) - E\mu(x_1)] + z_k$$

are total disturbances of the first and second Hammerstein systems, respectively. The measurement equations are as follows

$$X_N = \Omega_N q + E_N, \qquad Y_N = \Phi_N p + D_N \tag{2.99}$$

where

$$Y_N = (y_1, y_2, ..., y_N)^T, \tag{2.100}$$
$$\Phi_N = (\phi_1, \phi_2, ..., \phi_N)^T, \tag{2.101}$$
$$D_N = (d_1, d_2..., d_N)^T, \tag{2.102}$$
$$X_N = (x_1, x_2, ..., x_N)^T, \tag{2.103}$$
$$\Omega_N = (\varphi_1, \varphi_2, ..., \varphi_N)^T, \tag{2.104}$$
$$E_N = (e_1, e_2..., e_N)^T. \tag{2.105}$$

According to Assumption 2.25 and the fact that $E[\eta(u_{k-i}) - E\eta(u_1)] = 0$ and $E[\mu(x_{k-i}) - E\mu(x_1)] = 0$ it holds that

$$Ee_k = 0, \text{ and } Ed_k = 0 \text{ for } k = 1, 2, ..., N. \tag{2.106}$$

Standard least squares methodology applied to (2.99) leads to the following estimates of q and p

$$\hat{q}_N^{(LS)} = \left(\frac{1}{N}\Omega_N^T\Omega_N\right)^{-1} \frac{1}{N}\Omega_N^T X_N, \tag{2.107}$$

$$\hat{p}_N^{(LS)} = \left(\frac{1}{N}\Phi_N^T\Phi_N\right)^{-1} \frac{1}{N}\Phi_N^T Y_N. \tag{2.108}$$

Inserting (2.99) into (2.107) and (2.108) we obtain that $\widehat{q}_N^{(LS)} = q + \delta_N^{(LS)}$, and $\widehat{p}_N^{(LS)} = p + \Delta_N^{(LS)}$, where

$$\delta_N^{(LS)} = \left(\frac{1}{N} \Omega_N^T \Omega_N \right)^{-1} \frac{1}{N} \Omega_N^T E_N, \qquad (2.109)$$

$$\Delta_N^{(LS)} = \left(\frac{1}{N} \Phi_N^T \Phi_N \right)^{-1} \frac{1}{N} \Phi_N^T D_N \qquad (2.110)$$

are estimation errors. First of them, $\delta_N^{(LS)}$, obviously tends to zero, as $N \to \infty$, since the elements e_k of the vector E_N are independent of the input u_k included in φ_k. Nevertheless, the estimation error of second system nonlinearity, $\Delta_N^{(LS)}$, does not tend to zero, since the elements d_k of the vector D_N are correlated with the interaction input x_k included in ϕ_k. Instead of least squares, we thus propose the instrumental variables estimate of the form

$$\widehat{p}_N^{(IV)} = \left(\frac{1}{N} \Psi_N^T \Phi_N \right)^{-1} \frac{1}{N} \Psi_N^T Y_N, \qquad (2.111)$$

and postulate fulfillment of conditions *(C1)* and *(C2)* (see page 63).

2.5.4 *Nonparametric Generation of Instruments*

Obviously, the properties of the internal signal $\eta(u_k)$ of the first Hammerstein system allows to build the matrix of instruments Ψ_N for the second system, such that *(C1)* and *(C2)* are fulfilled. Here we solve that problem assuming temporarily that $\eta(u_k)$ is known. Next, in Section 2.5.4 we apply nonparametric (kernel or orthogonal) estimate of the regression function in the first Hammerstein system, which plays the role of deconvolution of the process x_k. The obtained estimates $\widehat{\eta}_M(u_k)$, computed from M observations $\{(u_k, x_k)\}_{k=1}^M$ are plugged in to the vectors of N instruments ψ_k $(k = 1, 2, ..., N)$ in the parametric identification procedure of second system. The relation between N and M, which guarantees consistency of the combined parametric-nonparametric estimate, is also given.

Known $\{\eta(u_k)\}$

Let us start from the simplest case, in which the internal signal (nonlinearity output) $\eta(u_k)$ of the first Hammerstein system can be observed and let the linear dynamic block $\{\lambda_i\}_{i=0}^\infty$ remains unknown. The following theorem holds.

Theorem 2.11. *The instrumental variable matrix Ψ_N generated according to the rule*

$$\overline{\Psi}_N = (\psi_1, \psi_2, ..., \psi_N)^T, \text{ where } \overline{\psi}_k = \left(1, \eta(u_k), \eta^2(u_k), ..., \eta^{W-1}(u_k)\right)^T,$$

$$(2.112)$$

fulfils (C1) *and* (C2).

Proof. For the proof see Appendix A.13.   ∎

Unknown $\{\eta(u_k)\}$

Theorem 2.11 guarantees that the estimate $\widehat{p}_N^{(IV)} = \left(\frac{1}{N}\overline{\Psi}_N^T \Phi_N\right)^{-1} \frac{1}{N}\overline{\Psi}_N^T Y_N$ is strongly consistent, i.e., converges with probability 1 to the parameter vector p of the identified function $\mu(x)$. However, the process $\{\eta(u_k)\}$, appearing in the instruments $\overline{\psi}_k$, cannot be measured and its estimation is needed. Similarly as in [98] we propose the preliminary deconvolution of $\{x_k\}$ to provide estimates $\widehat{\eta}_M(u_k)$ of $\eta(u_k)$, but in the problem considered here, we can apply nonparametric regression estimation methods. The inverse filtering of $\{x_k\}$ is made by the standard kernel or orthogonal algorithms of the regression function estimation in the first Hammerstein system, using the pairs $\{(u_k, x_k)\}_{k=1}^M$. We emphasize that $\widehat{\eta}_M(u_k)$'s are computed using the sequence of M measurements, which in general is greater than N, used in the parametric (instrumental variables) stage. The following theorem shows the relation between N and M, which guarantees the consistency of the two-stage estimate.

Theorem 2.12. *If the estimation error of* $\widehat{\eta}_M(u_k)$ *behaves like*

$$|\widehat{\eta}_M(u_k) - \eta(u_k)| = O(M^{-\tau}) \text{ with probability 1 as } M \to \infty, \qquad (2.113)$$

then for the instruments

$$\widehat{\Psi}_{N,M} = (\widehat{\psi}_{1,M}, \widehat{\psi}_{2,M}, ..., \widehat{\psi}_{N,M})^T, \qquad (2.114)$$

$$\text{where } \widehat{\psi}_{k,M} = \left(1, \widehat{\eta}_M(u_k), \widehat{\eta}_M^2(u_k), ..., \widehat{\eta}_M^{W-1}(u_k)\right)^T,$$

the two-stage estimate

$$\widehat{p}_{N,M}^{(IV)} = \left(\frac{1}{N}\widehat{\Psi}_{N,M}^T \Phi_N\right)^{-1} \frac{1}{N}\widehat{\Psi}_{N,M}^T Y_N \qquad (2.115)$$

is strongly consistent, i.e.,

$$\widehat{p}_{N,M}^{(IV)} \to p \text{ with probability 1 as } N, M \to \infty,$$

provided that $NM^{-\tau} \to 0$.

Proof. From (2.113) we conclude that for all $w = 1, 2, ..., W$ and $k = 1, 2, ..., N$, it holds that $|\widehat{\eta}_M^w(u_k) - \eta^w(u_k)| = O(M^{-w\tau})$, and consequently $\left\|\widehat{\psi}_{k,M} - \overline{\psi}_k\right\| = O(M^{-w\tau})$ with probability 1, as $M \to \infty$, where $\overline{\psi}_k$ is given by (2.112). Now, Theorem 2.12 may be proved in a similar manner as Theorem 3 in [65], with obvious substitutions, and hence the proof is here omitted. ∎

The Remark below, gives constructive recommendation for the selection of τ.

Remark 2.15. *[50] If $\eta()$ and the input probability density $f()$ are at least two times continuously differentiable at u, then for $h_M = h_0 M^{-1/5}$ the convergence rate is*

$$|\widehat{\eta}_M(u) - \eta(u)| = O(M^{-2/5})$$

in probability, i.e., $\tau = 2/5$ in Theorem 2.12.

2.5.5 Computer Experiment

The algorithm (2.107)-(2.108) was tested on the simple example of cascade connection of two identical Hammerstein models. We simulated the systems with odd polynomial characteristics

$$\eta(u) = u^3, \qquad \mu(x) = x^3, \tag{2.116}$$

followed by the IIR linear dynamic blocks

$$\lambda_i = \gamma_i = \left(\frac{1}{2}\right)^i, \qquad i = 0, 1, 2, \dots \tag{2.117}$$

excited by the uniformly distributed random input

$$u_k \sim U[-1, 1], \tag{2.118}$$

and the random disturbances

$$\varepsilon_k \sim U[-0.1, 0.1], \qquad z_k \sim U[-0.1, 0.1]. \tag{2.119}$$

From (2.116), (2.118), and (2.119) we simply conclude that $E\eta(u_1) = 0$ and $E\mu(x_1) = 0$, and hence (see (2.97))

$$m = 0 \text{ and } l = 0,$$

i.e., the problem of offsets in the semi-parametric approach is here hidden. In the identification stage of the experiment we only assumed the knowledge of orders of true polynomials $\eta()$ and $\mu()$ ($W \leqslant 5$) and that the linear dynamic blocks included in the systems are stable. The experiment was carried out on $N = 200, 250, \dots, 3000$ simulated measurements $\{(u_k, x_k, y_k)\}_{k=1}^N$ with $M = \lfloor 0.0002 \cdot N^2 \rfloor$, and repeated $R = 10$ times for each N. Estimation errors were computed according to the following rules

$$ERR_q(N) = \frac{1}{R} \sum_{r=1}^{R} \left\| \widehat{q}_N^{(r)} - q \right\|_2, \qquad ERR_p(N) = \frac{1}{R} \sum_{r=1}^{R} \left\| \widehat{p}_N^{(r)} - p \right\|_2,$$

where $\widehat{q}_N^{(r)}$ and $\widehat{p}_N^{(r)}$ denote the realization of the least squares estimates (2.107) and (2.108), obtained in the rth re-run. As can be seen in Fig. 2.17, the estimation error of parameters of the second systems is biased, even asymptotically.

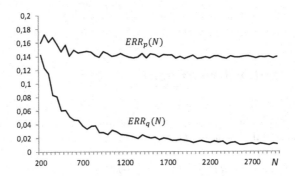

Fig. 2.17 Estimation error of the semi-parametric least squares method

2.5.6 Summary

It was shown, that instrumental variables algorithm proposed in Section 2.4 and [98] for identification of Hammerstein system with correlated input can be successfully applied, after slight modification, for cascade connection of two Hammerstein systems. Since the output of the first Hammerstein system is in fact a linear process, it is possible to perform its inverse filtering with the use of standard regression estimation methods. It is worth to notice that deconvolution of the interconnecting signal is made without any prior knowledge of the first Hammerstein system. Instrumental variables are thus generated in nonparametric way and support parameter estimate of the complementary block in the sense that its consistency is guaranteed.

2.6 Identification under Small Number of Data

In the nonparametric stage of the identification procedure the Hammerstein system is treated as the static system with the specific disturbance produced by the linear dynamic component. Since the dynamics is neglected the variance of the estimation error is huge for small number of data. The problem of data pre-filtering for nonparametric identification nonlinearity in Hammerstein system from short (finite) data set is considered in this section. The preliminary step is proposed. First, the linear dynamic block is approximated using instrumental variables technique, and the inverse of the obtained model is used for rough output filtering. Next, the standard procedure of non-parametric smoothing (kernel-based, or using orthogonal series expansion) is applied, involving the filtered output sequence instead of the original one. It is shown, that for small and moderate number of data, the estimation error can be significantly reduced in comparison with standard nonparametric methods. The asymptotic properties of the method (consistency and rate of convergence) remain the same as in the classical versions of nonparametric algorithms.

2.6.1 Introduction

When the nonlinear characteristic is recovered by the purely nonparametric method, the Hammerstein system is treated as a static system with specific disturbance and properties of the linear dynamics are, in some sense, ignored. In spite of their simplicity and good limit properties, the nonparametric estimates in standard versions are however inefficient for small and moderate number of data. The aim in this section is to propose the modified versions of nonparametric routines, more effective in the use with short data sequence (see [92]). The inverse filtering approach (see e.g. [17][31][138]) is used for improving the performance of the nonparametric estimate of the Hammerstein system nonlinearity. In particular, it is shown that for a finite number of data, the variance of the nonparametric estimate can be significantly reduced without affecting its limit properties. The two stage algorithm for nonlinearity recovering, exploiting the new idea of the combined approach to system identification, is presented. The parameter form of the static characteristic is not known (in the contrary to [64] and [65]) and the input signal is assumed to be a random process (in contrast to e.g. [75]). The strategy allows to incorporate additional knowledge about the linear dynamic block to improve small sample size properties of the nonparametric estimates of the static characteristic.

2.6.2 Statement of the Problem

Assumptions

In this section we assume that:

Assumption 2.28. *The nonlinear characteristic $\mu(u)$ is a Lipschitz function, i.e., it exists a positive constant $l < \infty$, such that for each $u, x \in R$ it holds $|\mu(u) - \mu(x)| \leqslant l\,|u - x|$.*

Assumption 2.29. *The linear dynamics with the unknown impulse response $\{\gamma_i\}_{i=0}^{\infty}$ is stable, i.e. $\sum_{i=0}^{\infty} |\gamma_i| < \infty$, and can be described by the AR(p) difference equation, i.e., $v_k = \sum_{i=1}^{p} a_i v_{k-i} + b_0 w_k$. The order p is finite and known a priori.*

Assumption 2.30. *The input $\{u_k\}$ and the noise $\{z_k\}$ are mutually independent i.i.d. random processes, $\sigma_u^2 = \mathrm{var}u_k < \infty$, $\sigma_z^2 = \mathrm{var}z_k < \infty$ and $Ez_k = 0$. It exists the input probability density $f(u)$.*

Similarly as in the previous sections, the objective is to recover $\mu(u)$ using input-output measurements $\{(u_k, y_k)\}_{k=1}^N$ of the *whole* Hammerstein system.

2.6.3 The Proposed Algorithm

As was mentioned above, the standard nonparametric methods ([40]-[50]) for recovering of the regression function

$$R(u) \triangleq E\{y_k|u_k = u\} = \gamma_0\mu(u) + \delta, \tag{2.120}$$

work completely independently of the shape of impulse response of the linear dynamics. The cost paid for simplicity, robustness and universality is a high variance of estimates. In the standard approach the Hammerstein system is treated in fact as a nonlinear static element corrupted by a correlated noise. Namely, one can specify three components of the output, i.e.,

$$y_k = \mu(u_k) + \sum_{i=1}^{\infty} \gamma_i\mu(u_{k-i}) + z_k. \tag{2.121}$$

The most part of the signal y_k is in a sense wasted, because the "system noise" $\xi_k \triangleq \sum_{i=1}^{\infty} \gamma_i\mu(u_{k-i})$ produced by linear dynamics is treated as an additional disturbance.

Under Assumption 2.29 and the fact that $v_k = y_k - z_k$ we obtain that

$$y_k = a_1y_{k-1} + a_2y_{k-2} + ... + a_py_{k-p} + w_k + \tilde{z}_k \tag{2.122}$$
$$= a_1y_{k-1} + ... + a_py_{k-p} + Ew_k + \tilde{w}_k + \tilde{z}_k,$$

where $\tilde{w}_k \triangleq w_k - Ew_k$, and $\tilde{z}_k \triangleq z_k - a_1z_{k-1} - a_2z_{k-2} - ... - a_pz_{k-p}$ are zero-mean stationary random processes. Equation (2.122) may be rewritten in the form

$$y_k = a_1y_{k-1} + a_2y_{k-2} + ... + a_py_{k-p} + c + d_k, \tag{2.123}$$

where $c = Ew_k$ is some (not informative) constant, and

$$d_k = \tilde{z}_k + \tilde{w}_k \tag{2.124}$$

may be interpreted as a zero-mean correlated random disturbance. Introducing the regressor $\phi_k = (y_{k-1}, y_{k-2}, ..., y_{k-p}, 1)^T \in R^{(p+1)\times 1}$, and the vector of unknown parameters

$$\theta = (a_1, a_2, ..., a_p, c)^T \in R^{(p+1)\times 1}, \tag{2.125}$$

we get $y_k = \phi_k^T \theta + d_k$, $k = 1, 2, ..., N$, or equivalently, in the compact, matrix-vector version $Y_N = \Phi_N \theta + D_N$, where $Y_N = (y_1, y_2, ..., y_N)^T \in R^{N \times 1}$, $\Phi_N = (\phi_1, \phi_2, ..., \phi_N)^T \in R^{N \times (p+1)}$, and $D_N = (d_1, d_2, ..., d_N)^T \in R^{N \times 1}$.

The Two Stage Method

The scheme of the proposed procedure is presented below.

Step 1. Identify the parameter vector (2.125) by the instrumental variables method

$$\widehat{\theta}_N = (\widehat{a}_{1,N}, \widehat{a}_{2,N}, ..., \widehat{a}_{p,N}, \widehat{c}_N)^T \qquad (2.126)$$
$$= (\Psi_N^T \Phi_N)^{-1} \Psi_N^T Y_N,$$
$$\text{where } \Psi_N = (\psi_1, \psi_2, ..., \psi_N)^T \in R^{N \times (p+1)},$$
$$\psi_k = (\psi_{k,1}, \psi_{k,2}, ..., \psi_{k,p}, \psi_{k,p+1})^T \in R^{(p+1) \times 1}$$

is additional matrix including instruments ψ_k, which fulfill the following two standard conditions [65]

$$\det E\psi_k \phi_k^T \neq 0 \text{ and } E\psi_k d_k = 0, \qquad (2.127)$$

and perform the following FIR output pre-filtering

$$y_k^f = y_k - \widehat{a}_{1,N} y_{k-1} - \widehat{a}_{2,N} y_{k-2} - ... - \widehat{a}_{p,N} y_{k-p}. \qquad (2.128)$$

Step 2. Using the filtered data $\{u_k, y_k^f\}_{k=1}^N$, compute the nonparametric estimate

$$\widehat{\mu}_N^f(u) = \frac{\sum_{k=1}^N y_k^f K\left(\frac{u_k - u}{h_N}\right)}{\sum_{k=1}^N K\left(\frac{u_k - u}{h_N}\right)}, \qquad (2.129)$$

or

$$\widehat{\mu}_N^f(u) = \frac{\sum_{i=0}^{q(N)} \widehat{\alpha}_{i,N}^f \varphi_i(u)}{\sum_{i=0}^{q(N)} \widehat{\beta}_{i,N} \varphi_i(u)}, \text{ where } \widehat{\alpha}_{i,N}^f = \frac{1}{N} \sum_{k=1}^N y_k^f \varphi_i(u_k). \qquad (2.130)$$

Remark 2.16. *Step 1 of the procedure can be replaced by any deconvolution method [17], and generalized for a class of invertible ARMA models. We assume AR dynamics and present the instrumental variables method, because each invertible filter can be approximated with arbitrarily small error by AR(p) model, when p grows large [15].*

Remark 2.17. *Conditions (2.127) mean that the instruments ψ_k should be correlated with output and simultaneously not correlated with the noise. We refer the reader to [65], where the universal method of generation of instruments for Hammerstein system is introduced. Here we assume, for shortness, that we know a priori the function m(), such that $m \triangleq E\mu(u_k)m(u_k) \neq 0$. In particular, we can choose $m(u) = u, u^3, ...$ if $\mu()$ is odd, or $m(u) = |u|, u^2$,*

... if $\mu()$ is even (see Remark 1 in [48]). Below, we show that the instruments of the form

$$\psi_k = (m(u_{k-1}), m(u_{k-2}), ..., m(u_{k-p}), 1)^T$$

fulfill (2.127).

Limit Properties

The following theorem holds.

Theorem 2.13. *If $|\widehat{\mu}_N(u) - \mu(u)| = O(N^{-\tau})$, $0 < \tau < 1/2$, in probability as $N \to \infty$, then*

$$\left|\widehat{\mu}_N^f(u) - \mu(u)\right| = O(N^{-\tau}) \qquad (2.131)$$

in probability as $N \to \infty$.

Proof. For the proof see Appendix A.14.   ∎

The Properties for $N < \infty$

Owing to (2.121) and (A.25) in Appendix A.14 we have respectively $y_k = w_k + \delta_k$, and $y_k^f = w_k + \delta_k^f$, where $\delta_k \triangleq \sum_{i=1}^{\infty} \gamma_i w_{k-i} + z_k$, and $\delta_k^f \triangleq \widetilde{z}_k + \sum_{i=1}^{p} (a_i - \widehat{a}_{i,N}) y_{k-i}$. One can show that the variance of the nonparametric regression function estimate is bounded from above as follows

$$\text{var}\widehat{\mu}_N(u) \leqslant c_0 n(N) \text{var}\delta_k, \qquad (2.132)$$

where c_0 is some constant dependent of the used kernel or basis functions, and $n(N) = \frac{1}{N h_N}$ for kernel estimate or $n(N) = \frac{q(N)}{N}$ for orthogonal methods (for details see e.g. Appendices I-II in [69], and Section 6 in [49]). The variances of δ_k and δ_k^f which influence the upper bounds (see (2.132)) of the variances of (1.27), (1.36), (2.129) and (2.130) have the form

$$\text{var}\delta_k = \sigma_z^2 + \sigma_w^2 \sum_{i=1}^{\infty} \gamma_i^2, \qquad (2.133)$$

$$\text{var}\delta_k^f = \text{var}\left(\widetilde{z}_k + \sum_{i=1}^{p} (a_i - \widehat{a}_{i,N}) y_{k-i}\right) \leqslant \qquad (2.134)$$

$$\leqslant \sigma_z^2 \left(1 + \sum_{i=1}^{p} a_i^2\right) + \frac{c_v}{N},$$

where c_v is some constant. Equations (2.133) and (2.134) illustrate the effect of filtration. If $\sigma_z^2 \left(1 + \sum_{i=1}^{p} a_i^2\right) < \sigma_z^2 + \sigma_w^2 \sum_{i=1}^{\infty} \gamma_i^2$, which is often the case in Hammerstein system, then it exists N_0 such that for all $N > N_0$ it holds that $\text{var}\delta_k^f < \text{var}\delta_k$. The variance of measurement noise is slightly amplified, but

Table 2.3 *MISE* of the kernel / orthogonal expansion estimates

N	(1.27)	(2.129)	(1.36)	(2.130)
50	37.37	2.81	41.90	3.93
100	27.11	2.18	34.19	3.01
300	15.48	0.89	17.30	2.81
500	9.77	0.57	12.12	2.67
1000	7.33	0.46	9.06	2.20
3000	4.42	0.22	4.87	1.25

the method gets rid of the harmful influence of the dynamics. For example, if $\sigma_z^2 = 1$, $\sigma_w^2 = 100$, $p = 1$, and $a_1 = 0.8$, then $\mathrm{var}\delta_k = 401$ whereas $\mathrm{var}\delta_k^f \simeq 1.64$.

2.6.4 Numerical Example

The input u_k and the noise z_k are uniformly distributed on $[-2\pi, 2\pi]$ and $[-0.1, 0.1]$, respectively. We took the AR(1) linear dynamics: $v_k = 0.8v_{k-1} + w_k$, and the nonlinear characteristic:

$$w_k = \mu(u_k) = \left(\frac{1}{2} + \mathrm{sgn}(u_k) \right) (||u_k| - \pi| - \pi) + 2\sin u_k.$$

In Step 1 we set $\psi_k = (u_{k-1}, u_{k-2}, ..., u_{k-p}, 1)^T$. In the kernel-type estimation algorithm (2.129) we applied the window kernel $K(x) = \begin{cases} 1, & \text{as } |x| < 1 \\ 0, & \text{elsewhere} \end{cases}$, and set $h(N) \sim N^{-1/5}$. In the orthogonal series expansion method (2.130) we used trigonometric orthonormal system

$$\frac{1}{\sqrt{4\pi}}, \frac{1}{\sqrt{2\pi}}\cos\frac{u}{2}, \frac{1}{\sqrt{2\pi}}\sin\frac{u}{2}, \frac{1}{\sqrt{2\pi}}\cos\frac{2u}{2}, \frac{1}{\sqrt{2\pi}}\sin\frac{2u}{2}, \frac{1}{\sqrt{2\pi}}\cos\frac{3u}{2}, ...$$

and set $q(N) \sim N^{1/5}$. Both methods have been compared with their classical versions (1.27) and (1.36). The mean integrated squared error has been computed numerically, according the rule $MISE\widehat{e}_N(u) = \int_{-2\pi}^{2\pi} (\widehat{e}_N(u) - \mu(u))^2 \, du$, where $\widehat{e}_N(u)$ stands for $\widehat{\mu}_N^f(u)$ and $\widehat{\mu}_N(u)$ respectively. The results are presented in Table 3.3. For small and moderate sample sizes the estimation error has been reduced about 15 times.

2.6.5 Conclusions

Additional prior knowledge about the linear dynamic block allows to speed up the convergence of nonparametric regression-type estimates of nonlinearity

in Hammerstein system. The proper output filtering reduces the estimation
error for small and moderate number of measurements, and does not affect
the limit properties even if the assumed model of the linear dynamics is not
correct. In the light of this, nonparametric methods with data pre-filtering
are worth further studies.

2.7   Semiparametric Approach

The idea of semiparametric approach to Hammerstein system identification is proposed e.g. in ([46]) and ([126]). The estimates are obtained by
incorporating a parametric component into the kernel, orthogonal or
correlation-based nonparametric algorithm. In this section we present two
semiparametric algorithms to recover a nonlinear characteristic in a Hammerstein system. For small number of observations, their identification errors are smaller than that of the purely nonparametric algorithm. The same
idea is also proposed for identification of linear dynamic component. Purely
parametric instrumental variables estimate is elastically substituted by the
nonparametric (correlation-based) method, when the number of observations
tends to infinity.

2.7.1   *Nonlinearity Recovering*

Let $\rho^* u$ be the best linear approximation ρu of $\mu(u)$, i.e., ρu with ρ minimizing
$E\left(\mu(u_k) - \rho u_k\right)^2$. The optimal approximation is with $\rho^* = E\left\{y_k u_k\right\}/E u_k^2$,
which can be estimated in the following way

$$\hat{\rho} = \frac{\frac{1}{N}\sum\limits_{k=1}^{N} y_k u_k}{\sum\limits_{k=1}^{N} u_k^2}.$$

We define

$$\hat{p}(u) = \hat{\rho}_N u \tag{2.135}$$

as an estimate of $\rho^* u$. The remaining part of $\mu(u)$, i.e., $q(u) = \mu(u) - \rho^* u$,
we estimate with the kernel estimate, see (1.27), in the following way

$$\hat{q}(u) = \frac{\sum\limits_{k=1}^{N} (y_k - \hat{\rho} u_k) K\left(\dfrac{u - u_k}{h_N}\right)}{\sum\limits_{k=1}^{N} K\left(\dfrac{u - u_k}{h_N}\right)}.$$

Therefore, our first semiparametric estimate has the following form:

$$\hat{\mu}_1(u) = \hat{p}(u) + \hat{q}(u). \tag{2.136}$$

Its first component recovers $\rho^* u$ in a parametric way, the other estimates the remaining part $q(u) = \mu(u) - \rho^* u$ in a nonparametric way. Advantages of the first dominate for small N, of the other for large. From the property that

$$\hat{\mu}(u) \to \mu(u), \text{ as } N \to \infty \text{ in probability,} \tag{2.137}$$

it follows that $\hat{q}(u) \to q(u)$ as $N \to \infty$ in probability at every point u at which both $\mu()$ and the input probability density function $f()$ are continuous and, moreover, $f(u) > 0$. Since $\hat{p}(u) \to \rho^* u$ as $N \to \infty$ in probability at every point u, our semiparametric estimate converges to the nonlinear characteristic, i.e.,

$$\hat{\mu}_1(u) \to \mu(u) \text{ as } N \to \infty \text{ in probability}$$

at every point u at which both $\mu()$ and $f()$ are continuous and $f(u) > 0$.

Our second semiparametric algorithm is of the following form

$$\hat{\mu}_2(u) = \lambda_N \hat{p}(u) + (1 - \lambda_N)\, \hat{\mu}(u), \tag{2.138}$$

with $\lambda_N \to 0$ as $N \to \infty$. It is just a linear combination of $\hat{p}(u)$ recovering the linear part and $\hat{\mu}(u)$ estimating $\mu(u)$. It is obvious that consistency of $\hat{\mu}(u)$ implies consistency of $\hat{\mu}_2(u)$, which means that

$$\hat{\mu}_2(u) \to \mu(u) \text{ as } N \to \infty \text{ in probability}$$

at every point u at which both $\mu()$ and $f()$ are continuous and $f(u) > 0$. If the balance in the combination is proper, for small N, MISE should decrease at the rate typical for parametric inference.

It is clear that the parametric component $\hat{p}(u)$ in (2.136) converges to $\rho^* u$ in probability at the rate $N^{-1/2}$, while the nonparametric one $\hat{q}(u)$ to $q(u)$ at the rate typical for nonparametric inference being slower than $N^{-1/2}$. The same is with (2.138), $\hat{p}(u)$ converges to $\rho^* u$ faster than $\hat{\mu}(u)$ to $\mu(u)$. Therefore for small N, where parametric components dominate nonparametric and errors of both (2.136) and (2.138) are expected to be smaller than that of (1.27).

2.7.2 Identification of the Linear Dynamic Subsystem

In general, the linear dynamic component with the impulse response $\{\gamma_j\}_{j=1}^{\infty}$, i.e.,

$$v_k = \sum_{j=1}^{\infty} \gamma_j w_{k-j}, \qquad \gamma_0 = 1 \tag{2.139}$$

cannot be described by the model, which includes finite number of parameters. Therefore, almost each purely parametric method is characterized by systematic approximation error. We will show that this error can be eliminated with the help of additional nonparametric term in the estimate. Let us consider the class of 2nd-order models with the transmittances

$$K(z, a_1, a_2) = \frac{z}{z^2 - a_1 z - a_2}, \qquad a_1, a_2 \in R \qquad (2.140)$$

and let the vector $a^* = (a_1^*, a_2^*)^T$ includes parameters of the best approximation of the true dynamic object (2.139) in the class (2.140), i.e.,

$$a_1^*, a_2^* = \arg \min_{a_1, a_2} E \left(y_k - a_1 y_{k-1} - a_2 y_{k-2} \right)^2$$

and let

$$\{\gamma_j^*\} = \mathcal{Z}^{-1} \left(K(z, a_1^*, a_2^*) \right)$$

be respective impulse response of this model. Traditionally, parameters a_1^*, and a_2^* are identified by the instrumental variables (IV) method (see [154] and [129])

$$\widehat{a}^{()} = \left(\widehat{a}_1^{(IV)}, \widehat{a}_2^{(IV)} \right) = \left(\Psi_N^T \Phi_N \right)^{-1} \Psi_N^T Y_N, \qquad (2.141)$$

where $\Phi_N = (\phi_1, \phi_2, ..., \phi_N)^T$, $\phi_k = (y_{k-1}, y_{k-2})^T$, $Y_N = (y_1, y_2, ..., y_N)^T$ and $\Psi_N = (\psi_1, \psi_2, ..., \psi_N)^T$ is the matrix of appropriately selected instruments, such that the following two conditions are fulfilled:

(C1) $\mathrm{Plim}\Psi_N^T \Phi_N$ exists and is not singular,

(C2) $\mathrm{Plim}\Psi_N^T \widetilde{Z}_N = 0$, where $\widetilde{Z}_N = (\widetilde{z}_1, \widetilde{z}_2, ..., \widetilde{z}_N)^T$ and $\widetilde{z}_k = z_k - a_1 z_{k-1} - a_2 z_{k-2}$.

The impulse response can be approximated indirectly as

$$\{\widehat{\gamma}_j^{(IV)}\} = \mathcal{Z}^{-1} \left(K(z, \widehat{a}_1^{(IV)}, \widehat{a}_2^{(IV)}) \right).$$

Since for $N \to \infty$ it holds that (see [65] and [154])

$$\widehat{a}^{(IV)} \to a^* \qquad \text{in probability,} \qquad (2.142)$$

under continuity of the transform $\mathcal{Z}^{-1}()$ we get also

$$\widehat{\gamma}_j^{(IV)} \to \gamma_j^* \qquad \text{in probability} \qquad (2.143)$$

for each $j = 1, 2, ...$ Let us however emphasize that $\gamma_j^* \neq \gamma_j$, whenever the parametric model is not correct. For N large, the true elements $\{\gamma_j\}$ of the impulse response of the linear dynamics can be recovered successfully by the nonparametric way, using input-output cross-correlation analysis ([48]). The nonparametric estimate has the form

$$\widehat{\gamma}_j^{(corr)} = \frac{\widehat{\varkappa}_j}{\widehat{\varkappa}_1}, \text{ where } \widehat{\varkappa}_j = \frac{1}{N-j} \sum_{k=1}^{N-j} y_{k+j} u_k. \tag{2.144}$$

Although the estimate (2.144) is consistent, i.e.,

$$\widehat{\gamma}_j^{(corr)} \to \gamma_j \qquad \text{in probability,}$$

as $N \to \infty$, it has huge variance-error, especially if $N - j$ is small. Therefore, analogously to (2.138) we propose to combine (2.126) with (2.144) as follows

$$\widehat{\gamma}_j^{(semi)} = \lambda_N \widehat{\gamma}_j^{(IV)} + (1 - \lambda_N)\widehat{\gamma}_j^{(corr)} \tag{2.145}$$

with $\lambda_N \to 0$ as $N \to \infty$.

2.8 Final Remarks

In the chapter we combined parametric and nonparametric methods in Hammerstein system identification. The global problem is decomposed, by nonparametric kernel regression estimate, to local identification tasks. Each block is identified independently of prior knowledge of the other subsystems. Our considerations admit broad class of nonlinearities, including functions which are not linear in the parameters, and any stable IIR linear dynamic blocks. Moreover, the algorithms cope with correlated excitations, thanks to nonparametric procedures of instrumental variables generation. Thus, cascade complex systems can be identified without asymptotic bias. Also the semiparametric approach is presented, which allows to speed up the convergence, i.e., to improve the results for finite number of observations.

Chapter 3
Wiener System

3.1 Introduction to Wiener System

Since in the Wiener system (Fig. 3.1) the nonlinear block is preceded by the
linear dynamics and the nonlinearity input is correlated, its identification is
much more difficult in comparison with the Hammerstein system. However,
the Wiener model allows for better approximation of many real processes. The
difficulties in theoretical analysis force the authors to consider special cases,
and to take somehow restrictive assumptions on the input signal, impulse
response of the linear dynamic block and the shape of the nonlinear characteristics. In particular, for Gaussian input the problem of Wiener system
identification becomes much easier. Since the internal signal $\{x_k\}$ is then also
Gaussian, the linear block can be simply identified by the cross-correlation
approach, and the static characteristic can be recovered e.g. by the nonparametric inverse regression approach ([41]-[42]) provided that it is locally
invertible. Non-Gaussian random input is very rarely met in the literature.
It is allowed e.g. in ([108]), but the algorithm presented there requires prior
knowledge of the parametric representation of the linear subsystem. Most of
recent methods for Wiener system identification assume FIR linear dynamics,
invertible nonlinearity, or require the use of specially designed input excitations ([10], [5], [55], [37], [8]). The most general assumptions were taken in
([90]), ([94]), ([44]) and ([97]), but the convergence of the proposed estimates
is slow. Similarly as for Hammerstein system, in this chapter we compare
and combine two kinds of methods, parametric ([3], [5], [9]-[37], [55]-[81],
[111]-[146], [151], [153], [159]) and nonparametric ([41]-[71], [90]-[108], [147]).

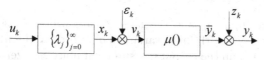

Fig. 3.1 Wiener system

G. Mzyk, *Combined Parametric-Nonparametric Identification*
of Block-Oriented Systems, Lecture Notes in Control and Information Sciences 454,
DOI: 10.1007/978-3-319-03596-3_3, © Springer International Publishing Switzerland 2014

The method is called 'parametric' if both linear and nonlinear components of Wiener system are described with the use of finite and known number of unknown parameters, e.g. when FIR linear dynamic model and polynomial characteristic are assumed. The popular parametric methods elaborated for Wiener system identification are not free of the approximation error and do not allow full decentralization of the identification task of complex system. Moreover the theoretical analysis of identifiability, and convergence of parametric estimates remains relatively difficult. On the other hand, nonparametric approach offers simple algorithms, which are asymptotically free of approximation error, i.e. they converge to the true system characteristics. However, the purely nonparametric methods are not commonly exploited in practice for the following reasons: (i) they depend on various tuning parameters and functions; in particular, proper selection of kernel and the bandwidth parameter or orthonormal basis and the scale factor are critical for the obtained results, (ii) the prior knowledge of subsystems is completely neglected; the estimates are based on measurements only, and the resulting model may be not satisfactory when the number of measurements is small, and (iii) bulk number of estimates must be computed when the model complexity grows large.

The traditional parametric nonlinear least-squares approach for Wiener system identification with its weak points was discussed in Section 1.4.2. In Section 3.2, we present several purely nonparametric methods, i.e., correlation-based estimate of the linear dynamics, kernel estimate of the inverse regression, and a censored sample mean approach to nonlinearity recovering. Finally, selected parametric and nonparametric methods are combined and the properties of the proposed two-stage procedures are discussed in Section 3.3.

3.2 Nonparametric Identification Tools

The nonparametric methods for a class of Wiener systems, with Gaussian input and locally invertible characteristics were introduced by Greblicki (see e.g. [42]). Respective algorithms are shortly reminded in Sections 3.2.1 and 3.2.2. In Section 3.2.3 we show a new kind of nonlinearity estimate in Wiener system, which works under the least possible prior knowledge, i.e., under non-Gaussian input, IIR linear dynamics and any continuous, but not necessary invertible, static characteristic.

3.2.1 Inverse Regression Approach

Assume that the input u_k and the noise ε_k are white Gaussian, mutually independent processes with finite variances $\sigma_u^2, \sigma_\varepsilon^2 < \infty$, the noise ε_k is zero-mean, i.e. $E\varepsilon_k = 0$, and the output measurement noise z_k is zero, i.e., accurate output $y_k = \bar{y}_k$ is observed. The nonparametric estimation of the inverted regression relies on the following lemma.

Lemma 3.1. *[50] If $\mu()$ is invertible then for any $y \in \mu(R)$ it holds that*

$$E\left(u_k | y_{k+p} = y\right) = \alpha_p \mu^{-1}(y) \tag{3.1}$$

where $\alpha_p = \lambda_p \frac{\sigma_u^2}{\sigma_v^2}$.

Since for any time lag p, the $\mu^{-1}(y)$ can be identified only up to some multiplicative constant α_p, let us denote, for convenience, $v(y) = \alpha_p \mu^{-1}(y)$. The nonparametric kernel estimate of $v(y)$ has the form

$$\widehat{v}(y) = \frac{\sum_{k=1}^{N} u_k K\left(\frac{y - y_{k+p}}{h(N)}\right)}{\sum_{k=1}^{N} K\left(\frac{y - y_{k+p}}{h(N)}\right)}, \tag{3.2}$$

where $K()$ and $h(N)$ is a kernel function and bandwidth parameter, respectively. Usability of (3.2) is based on the following theorem.

Theorem 3.1. *[50] If $\mu()$ is invertible, $K()$ is Lipschitz and such that $c_1 H(|y|) \leqslant K(y) \leqslant c_2 H(|y|)$ for some c_1 and c_2, where $H()$ is nonnegative and non-increasing function, defined of $[0, \infty)$, continuous and positive at $t = 0$, and such that $tH(t) \to 0$ as $t \to \infty$, then for $h(N) \to 0$ and $Nh^2(N) \to \infty$ as $N \to \infty$ it holds that*

$$\widehat{v}(y) \to v(y) \text{ in probability as } N \to \infty, \tag{3.3}$$

at every point y, in which the probability density $f(y)$ is positive and continuous.

The rate of convergence in (3.3) depends on the smoothness of the identified characteristic and is provided by the following lemma.

Lemma 3.2. *[50] Let us define $g(y) \triangleq v(y)f(y)$ and denote $v(y) = \frac{g(y)}{f(y)}$. If $\mu^{-1}()$, $f()$ and $g()$ have q bounded derivatives in a neighborhood of y, then*

$$|\widehat{v}(y) - v(y)| = \mathcal{O}\left(N^{-\frac{1}{2} + \frac{1}{2q+2}}\right) \text{ in probability.}$$

e.g. $\mathcal{O}(N^{-1/4})$ for $q = 1$, $\mathcal{O}(N^{-1/3})$ for $q = 2$, and $\mathcal{O}(N^{-1/2})$ for q large.

In [42] and [43], the estimate (3.2) was generalized for the larger class of Wiener systems, admitting the "locally invertible" nonlinear static blocks and correlated excitation. The strongest limitation of the inverse regression approach is thus assumption about the Gaussianity of the input signal.

3.2.2   Cross-Correlation Analysis

The nonparametric identification of the linear dynamic block is based on the following property.

Lemma 3.3. *[50] If $E\,|v_k\mu(v_k)| < \infty$ then*

$$E\{u_k y_{k+p}\} = \beta\lambda_p,$$

where $\beta = \frac{\sigma_u^2}{\sigma_v^2} E\{v_k\mu(v_k)\}$.

Since λ_p can be identified only up to some multiplicative constant β, let us denote, for convenience, $\kappa_p \triangleq \beta\lambda_p$, and consider its natural estimate of the form

$$\widehat{\kappa}_p = \frac{1}{N}\sum_{k=1}^{N} u_k y_{k+p}. \tag{3.4}$$

Theorem 3.2. *[50] If $\mu()$ is the Lipschitz function, then*

$$\lim_{N\to\infty} E\left(\widehat{\kappa}_p - \kappa_p\right)^2 = \mathcal{O}\left(\frac{1}{N}\right) \tag{3.5}$$

Consequently, when the stable IIR linear subsystem is modelled by the filter with the impulse response $\widehat{\kappa}_0, \widehat{\kappa}_1, ..., \widehat{\kappa}_{n(N)}$, then it is free of the asymptotic approximation error if $n(N) \to \infty$ and $n(N)/N \to 0$ as $N \to \infty$.

3.2.3   A Censored Sample Mean Approach

The estimate presented in this section (see [90] and [93]) works under small amount of a priori information. An IIR dynamics, non-invertible static non-linearity, and non-Gaussian excitations are admitted. The convergence of the estimate is proved for each continuity point of the static characteristic and the asymptotic rate of convergence is analyzed. The results of computer simulation example are included to illustrate the behaviour of the estimate for moderate number of observations.

Introduction

The results presented below are the extension and generalization of the idea shown in [90], in the sense that:

- A new method of the identification of Wiener system nonlinearity is proposed, which works under mild assumptions on the subsystems and excitations. In particular, in contrast to most earlier papers: (1) the input sequence need not to be a Gaussian white noise, (2) the nonlinear characteristic is not assumed to be invertible, (3) the IIR linear dynamics is admitted, and (4) the algorithm is of nonparametric nature (see e.g. [50]), i.e. it is not assumed that the subsystems can be described with the use of finite number of parameters; in consequence the estimate is free of the possible approximation error. We provide strict convergence proof of our, kernel-based, estimate of the static characteristic, and, in addition to [90], analyze the asymptotic rate of convergence also for the case of IIR dynamic subsystem.

- The convergence proof is next generalized for Wiener-Hammerstein system, and Hammerstein system as a particular cases of the latter.
- We exploit the idea of the combined parametric-nonparametric approach (see [64], [65] and [92]) to system identification, and in this context we present the use of the proposed method as a preliminary step for parameter estimation of nonlinear subsystem, when its parametric description is a priori known, i.e., when we are given the closed formula describing the nonlinearity, which includes finite number of unknown parameters. This formula need not to be linear in the parameters.

Assumptions

We assume that:

Assumption 3.1. *The input $\{u_k\}$ is an i.i.d., bounded ($|u_k| < u_{\max}$; unknown $u_{\max} < \infty$) random process. There exists a probability density of the input, $\vartheta_u(u_k)$ say, which is a continuous and strictly positive function around the estimation point x, i.e., $\vartheta_u(x) \geqslant \varepsilon > 0$.*

Assumption 3.2. *The unknown impulse response $\{\lambda_j\}_{j=0}^{\infty}$ of the linear IIR filter is exponentially upper bounded, that is*

$$|\lambda_j| \leqslant c_1 \lambda^j, \text{ some unknown } 0 < c_1 < \infty, \qquad (3.6)$$

where $0 < \lambda < 1$ is an a priori known constant.

Assumption 3.3. *The nonlinearity $\mu(x)$ is an arbitrary function, continuous almost everywhere on $x \in (-u_{\max}, u_{\max})$ (in the sense of Lebesgue measure).*

Assumption 3.4. *The output noise $\{z_k\}$ is a zero-mean stationary and ergodic process, which is independent of the input $\{u_k\}$.*

Assumption 3.5. *For simplicity of presentation we also let $L \triangleq \sum_{j=0}^{\infty} \lambda_j = 1$ and $u_{\max} = \frac{1}{2}$.*

The goal is to estimate the unknown characteristic of the nonlinearity $\mu(x)$ on the interval $x \in (-u_{max}, u_{max})$ on the basis of M input-output measurements $\{(u_k, y_k)\}_{k=1}^{M}$ of the whole Wiener system.

Comments to Assumptions

1) We emphasize, that in Assumption 3.2, we do not assume parametric knowledge of the linear dynamics. In fact, the condition (3.6), with unknown c_1, is rather not restrictive, and characterizes the class of stable objects. Moreover, observe that, in particular case of FIR linear dynamics, Assumption 3.2 is fulfilled for arbitrarily small $\lambda > 0$.

2) Assumption 3.5 is of technical meaning only. We note that the members of the family of Wiener systems composed by series connection of linear filters with the impulse responses $\{\overline{\lambda}_j\} = \{\frac{\lambda_j}{c_2}\}_{j=0}^{\infty}$ and the nonlinearities $\overline{\mu}(x) = \mu(c_2 x)$ are, for $c_2 \neq 0$, indistinguishable from the input-output point of view. In consequence, from the input-output viewpoint, $\mu()$ can be recovered in general only up to some domain scaling factor c_2, independently of the applied identification method.

3) From Assumptions 3.1 and 3.2 it holds that $|x_k| < x_{max} < \infty$, where $x_{max} \triangleq u_{max} \sum_{j=0}^{\infty} |\lambda_j|$. Since $\sum_{j=0}^{\infty} |\lambda_j| \geqslant L$ and $L = 1$ (see Assumption 3.4), thus the support of the random variables x_k, i.e. $(-x_{max}, x_{max})$, is generally wider than the estimation interval $x \in (-u_{max}, u_{max})$. We also introduce and analyze the nonparametric estimate of the part of characteristic $\mu(x)$, for $x \in (-u_{max}, u_{max})$, and we expand the obtained results for $x \in (-x_{max}, x_{max})$, when the parametric knowledge of $\mu()$ is provided.

Background of the Approach

Let x be a chosen estimation point of $\mu(\cdot)$. For a given x let us define a "weighted distance" between the measurements $u_k, u_{k-1}, u_{k-2}, ..., u_1$ and x as

$$\delta_k(x) \triangleq \sum_{j=0}^{k-1} |u_{k-j} - x| \lambda^j = |u_k - x| \lambda^0 + |u_{k-1} - x| \lambda^1 + ...$$

$$... + |u_1 - x| \lambda^{k-1}, \tag{3.7}$$

i.e. $\delta_1(x) = |u_1 - x|$, $\delta_2(x) = |u_2 - x| + |u_1 - x| \lambda$, $\delta_3(x) = |u_3 - x| + |u_2 - x| \lambda + |u_1 - x| \lambda^2$, etc., which can be computed recursively as follows

$$\delta_k(x) = \lambda \delta_{k-1}(x) + |u_k - x|. \tag{3.8}$$

Making use of assumptions Assumptions 3.5 and 3.2 we obtain that

$$|x_k - x| = \left| \sum_{j=0}^{\infty} \lambda_j u_{k-j} - \sum_{j=0}^{\infty} \lambda_j x \right| = \left| \sum_{j=0}^{\infty} \lambda_j (u_{k-j} - x) \right| =$$

$$= \left| \sum_{j=0}^{k-1} \lambda_j (u_{k-j} - x) + \sum_{j=k}^{\infty} \lambda_j (u_{k-j} - x) \right| \leqslant$$

$$\leqslant \sum_{j=0}^{k-1} |\lambda_j| \, |u_{k-j} - x| + 2 u_{\max} \sum_{j=k}^{\infty} |\lambda_j| \leqslant$$

$$\leqslant \delta_k(x) + \frac{\lambda^k}{1 - \lambda} \triangleq \Delta_k(x). \tag{3.9}$$

Observe that if in turn

$$\Delta_k(x) \leqslant h(M), \tag{3.10}$$

then the true (but unknown) interaction input x_k is located close to x, provided that $h(M)$ (further, a calibration parameter) is small. The distance given in (3.9) may be easily computed as the point x and the data $u_k, u_{k-1}, u_{k-2}, ..., u_1$ are each time at ones disposal. In turn, the condition (3.10) selects k's for which the input sequences $\{u_k, u_{k-1}, u_{k-2}, ..., u_1\}$ are such that the true nonlinearity inputs $\{x_k\}$ surely belong to the neighborhood of the estimation point x with the radius $h(M)$. Let us also notice that asymptotically, as $k \to \infty$, it holds that

$$\delta_k(x) = \Delta_k(x), \tag{3.11}$$

with probability 1.

Proposition 3.1 *If, for each $j = 0, 1, ..., \infty$ and some $d > 0$, it holds that*

$$|u_{k-j} - x| \leqslant \frac{d}{\lambda^j}, \tag{3.12}$$

then

$$|x_k - x| \leqslant d \log_\lambda d + d \frac{1}{1 - \lambda}. \tag{3.13}$$

Proof. For the proof see Appendix A.15.   ∎

Kernel-Like Estimate for Estimation of the Static Characteristic

We propose the following nonparametric kernel-like estimate of the nonlinear characteristic $\mu()$ at the given point x, exploiting the distance $\delta_k(x)$ between x_k and x, and having the form

$$\widehat{\mu}_M(x) = \frac{\sum_{k=1}^{M} y_k \cdot K\left(\frac{\delta_k(x)}{h(M)}\right)}{\sum_{k=1}^{M} K\left(\frac{\delta_k(x)}{h(M)}\right)}, \tag{3.14}$$

where $K()$ is a window kernel function of the form

$$K(v) = \begin{cases} 1, & \text{as } |v| \leqslant 1 \\ 0, & \text{elsewhere} \end{cases}. \tag{3.15}$$

Since the estimate (3.14) is of the ratio form we treat the case $0/0$ as 0.

Limit Properties

The Convergence

The following theorem holds.

Theorem 3.3. *If $h(M) = d(M)\log_\lambda d(M)$, where $d(M) = M^{-\gamma(M)}$, and $\gamma(M) = \left(\log_{1/\lambda} M\right)^{-w}$, then for each $w \in \left(\frac{1}{2}, 1\right)$ the estimate (3.14) is consistent in the mean square sense, i.e., it holds that*

$$\lim_{M \to \infty} E\left(\widehat{\mu}_M(x) - \mu(x)\right)^2 = 0. \tag{3.16}$$

Proof. For the proof see Appendix A.16. ∎

The Rate of Convergence

To establish the asymptotic rate of convergence we additionally assume that:

Assumption 3.6. *The nonlinear characteristic $\mu(x)$ is a Lipschitz function, i.e., it exists a positive constant $l < \infty$, such that for each $x_a, x_b \in R$ it holds that $|\mu(x_a) - \mu(x_b)| \leqslant l\,|x_a - x_b|$.*

For a window kernel (3.15) we can rewrite (3.14) as $\widehat{\mu}_M(x) = \frac{1}{S_0} \sum_{i=1}^{S_0} y_{[i]}$, where $[i]$'s are indexes, for which $K\left(\frac{\delta_{[i]}(x)}{h(M)}\right) = 1$, and S_0 is a random number of selected output measurements. For each $y_{[i]}$, $i = 1, 2, ..., S_0$, respective $x_{[i]}$ is such that $\left|x_{[i]} - x\right| \leqslant h(M)$. On the basis of Assumption 3.6 we obtain

$$\left|\mu(x_{[i]}) - \mu(x)\right| \leqslant lh(M),$$

which for $Ez_k = 0$ (see Assumption 3.4) leads to

$$\left|\text{bias}\widehat{\mu}_M(x)\right| = \left|Ey_{[i]} - \mu(x)\right| = \left|E\mu(x_{[i]}) - \mu(x)\right| \leqslant lh(M),$$
$$\text{bias}^2\widehat{\mu}_M(x) = O\left(h^2(M)\right). \tag{3.17}$$

For the variance we have

$$
\operatorname{var}\widehat{\mu}_M(x) = \sum_{n=0}^{M} P(S_0 = n) \cdot \operatorname{var}\left(\widehat{\mu}_M(x)|S_0 = n\right) =
$$

$$
= \sum_{n=1}^{M} P(S_0 = n) \cdot \operatorname{var}\left(\frac{1}{n}\sum_{i=1}^{n} y_{[i]}\right).
$$

Since, under strong law of large numbers and Chebychev inequality, it holds that $\lim_{M\to\infty} P\left(S_0 > \alpha E S_0\right) = 1$ for each $0 < \alpha < 1$ (see [90]), we asymptotically obtain that

$$
\operatorname{var}\widehat{\mu}_M(x) = \sum_{n>\alpha E S_0} P(S_0 = n) \cdot \operatorname{var}\left(\frac{1}{n}\sum_{i=1}^{n} y_{[i]}\right) \tag{3.18}
$$

with probability 1. Taking into account that $y_{[i]} = \overline{y}_{[i]} + z_{[i]}$, where $\overline{y}_{[i]}$ and $z_{[i]}$ are independent random variables we obtain that

$$
\operatorname{var}\left(\frac{1}{n}\sum_{i=1}^{n} y_{[i]}\right) = \operatorname{var}\left(\frac{1}{n}\sum_{i=1}^{n} \overline{y}_{[i]}\right) + \operatorname{var}\left(\frac{1}{n}\sum_{i=1}^{n} z_{[i]}\right). \tag{3.19}
$$

Since the process $\{z_{[i]}\}$ is ergodic, under strong law of large numbers, it holds that

$$
\operatorname{var}\left(\frac{1}{n}\sum_{i=1}^{n} z_{[i]}\right) = O\left(\frac{1}{Mp(M)}\right) = O\left(\frac{1}{M}\right). \tag{3.20}
$$

The process $\{\overline{y}_{[i]}\}$ is in general not ergodic, but in consequence of (3.10) it has compact support $[\mu(x) - lh(M), \mu(x) + lh(M)]$ and the following inequality is fulfilled

$$
\operatorname{var}\left(\frac{1}{n}\sum_{i=1}^{n} \overline{y}_{[i]}\right) \leqslant \operatorname{var}\overline{y}_{[i]} \leqslant (2lh(M))^2. \tag{3.21}
$$

From (3.18), (3.19), (3.20) and (3.21) we conclude that

$$
\operatorname{var}\widehat{\mu}_M(x) = O(h^2(M)), \tag{3.22}
$$

which in view of (3.17) leads to

$$
|\widehat{\mu}_M(x) - \mu(x)| = O(h^2(M)) \tag{3.23}
$$

in the mean square sense. A relatively slow rate of convergence, guaranteed in a general case, for $h(M)$ as in Theorem 3.3, is a consequence of small amount of a priori information. Emphasize that for, e.g., often met in applications piecewise constant functions $\mu(x)$, it exists $M_0 < \infty$, such that $\mathrm{bias}^2 \widehat{\mu}_M(x) = 0$ and var $\left(\frac{1}{n} \sum_{i=1}^{n} \bar{y}_{[i]} \right) = 0$ for $M > M_0$, and consequently $|\widehat{\mu}_M(x) - \mu(x)| = O(\frac{1}{M})$ as $M \to \infty$ (see (3.20)).

3.3 Combined Parametric-Nonparametric Approach

As was mentioned in Chapter 2, the idea of combined parametric-nonparametric approach to system identification was introduced by Hasiewicz and Mzyk in [64], and continued in [65], [90], [92], [108], [147] and [67]. For the Wiener system, the algorithms decompose the complex system identification task on independent identification problems for each components. The decomposition is based on the estimation of the nonlinearity inputs x_k. Next, using the resulting pairs $(u_k, \widehat{x}_k)$ and $(\widehat{x}_k, y_k)$, both linear dynamic and static nonlinear subsystems are identified separately by e.g. the least squares or by the kernel method. In the contrary to Hammerstein system, where the interaction can be estimated directly by any regression estimation method, for Wiener system the situation is more complicated as $x_k = \sum_{j=0}^{\infty} \lambda_j u_{k-j}$, and the impulse response of the linear dynamics must be estimated first, to provide indirect estimates of x_k.

3.3.1 Kernel Method with the Correlation-Based Internal Signal Estimation (FIR)

Here we assume that the input u_k is white and Gaussian, the nonlinear characteristic $\mu()$ is bounded by polynomial of any finite order, $cov(u_1, y_1) \neq 0$, and the linear dynamics is FIR with known order s, i.e. $x_k = \sum_{j=0}^{s} \lambda_j u_{k-j}$.

Observe that $E\{y_k | x_k = x\} = \mu(x)$. Since the internal signal x_k cannot be measured, the following kernel regression estimate is proposed (see [147])

$$\widehat{\mu}(x) = \frac{\sum_{k=1}^{N} y_k K\left(\frac{x - \widehat{x}_k}{h(N)}\right)}{\sum_{k=1}^{N} K\left(\frac{x - \widehat{x}_k}{h(N)}\right)}, \tag{3.24}$$

where $\widehat{x}_k$ is indirect estimate of βx_k (i.e. scaled x_k)

$$\widehat{x}_k = \sum_{j=0}^{s} \widehat{\kappa}_j u_{k-j}$$

based on the input-output sample correlation (see (3.4)). The following theorem holds.

Theorem 3.4. *[147] If $K()$ is Lipschitz then $\widehat{\mu}(x) \to \mu(x/\beta)$ in probability as $N \to \infty$. Moreover, if both $\mu()$ and $K()$ are twice differentiable, then it holds that*

$$|\widehat{\mu}(x) - \mu(x/\beta)| = \mathcal{O}\left(N^{-\frac{2}{5}+\epsilon}\right) \tag{3.25}$$

for any small $\epsilon > 0$, provided that $h(N) \sim N^{-1/5}$.

In practise, due to assumed Gaussianity of excitations, the algorithm (3.25) is rather recommended for the tasks, in which the input process can be freely generated.

3.3.2 Identification of IIR Wiener Systems with Non-Gaussian Input

The censored kernel-type algorithm (3.14) is applied in this section to support estimation of parameters, when our prior knowledge about the system is large, and in particular, the parametric model of the characteristic is known.

Assume that we are given the class $\mu(x,c)$, such that $\mu(x) \subset \mu(x,c)$, where $c = (c_1, c_2, ..., c_m)^T$ and let us denote by $c^* = (c_1^*, c_2^*, ..., c_m^*)^T$ the vector of true parameters, i.e., $\mu(x,c^*) = \mu(x)$. Let moreover the function $\mu(x,c)$ be by assumption differentiable with respect to c, and the gradient $\nabla_c \mu(x,c)$ be bounded in some convex neighbourhood of c^* for each x. We assume that c^* is identifiable, i.e., there exists a sequence $x^{(1)}, x^{(2)}, ..., x^{(N_0)}$ of estimation points, such that

$$\mu(x^{(i)}, c) = \mu(x^{(i)}), \; i = 1, 2, ..., N_0 \implies c = c^*.$$

The proposed estimate has two steps.

Step 1. For the sequence $x^{(1)}, x^{(2)}, ..., x^{(N_0)}$ compute N_0 pairs

$$\left\{ \left(x^{(i)}, \widehat{\mu}_M(x^{(i)}) \right) \right\}_{i=1}^{N_0},$$

using the estimate (3.14).

Step 2. Perform the minimization of the cost-function

$$Q_{N_0,M}(c) = \sum_{i=1}^{N_0} \left(\widehat{\mu}_M(x^{(i)}) - \mu(x^{(i)}, c) \right)^2,$$

with respect to variable vector c, and take

$$\widehat{c}_{N_0,M} = \arg \min_c Q_{N_0,M}(c) \tag{3.26}$$

as the estimate of c^*.

Theorem 3.5. *Since in* Step 1 *(nonparametric) for the estimate (3.14) it holds that* $\widehat{\mu}_M(x^{(i)}) \to \mu(x^{(i)})$ *in probability as* $M \to \infty$ *for each* $i = 1, 2, ..., N_0,$ *thus*

$$\widehat{c}_{N_0,M} \to c^*$$

in probability, as $M \to \infty$.

Proof. See the proof of Theorem 1 in [65]. ∎

Simulation Example

In the computer experiment we generated uniformly distributed i.i.d. input sequence $u_k \sim U[-1, 1]$ and the output noise $z_k \sim U[-0.1, 0.1]$. We simulated the IIR linear dynamic subsystems $x_k = 0.5x_{k-1} + 0.5u_k$ and $\overline{y}_k = 0.5\overline{y}_{k-1} + 0.5v_k$, i.e. $\lambda_j = \gamma_j = 0.5^{j+1}$, $j = 0, 1, ..., \infty$, sandwiched with the non-invertible and not linear in the parameters static nonlinear characteristic $\mu(x_k) = c_1^* x_k + c_2^* + c_3^* \sin(c_4^* x_k)$, with $c_1^* = 1$, $c_2^* = 0$, $c_3^* = 0.2$ and $c_4^* = 2\pi$. The nonparametric estimate (3.14) was computed in the $N_0 = 21$ equispaced points $x^{(i)} = -1 + \frac{i-1}{10}$, $i = 1, 2, ..., N_0$. In Assumption 3.2 we took $\lambda = 0.8$. The estimation error was computed according the rule

$$ERR\left(\widehat{\mu}_M(x)\right) = \sum_{i=1}^{N_0} \left(\widehat{\mu}_M(x^{(i)}) - \mu(x^{(i)})\right)^2.$$

The result of estimation for $M = 300$ is shown in Fig. 3.2. The criterion in (3.26) was minimized with the use of classical Levenberg-Marquardt algorithm. Figure 3.3 illustrates the consistency property. In the experiment, the characteristic of the static block was changed for $\mu(x) = \sqrt[3]{x}$, which is not Lipschitz at $x = 0$. The effect of slower convergence in the neighbourhood of $x = 0$ can be seen in Fig. 3.4. Next, the routine was repeated for various values of the tuning parameter h. As can be seen in Fig. 3.5, according to intuition, improper selection of h results in variance or bias augmentation.

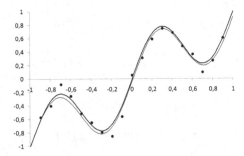

Fig. 3.2 The true characteristic $\mu(x) = x + 0.2 \sin 2\pi x$ (thick line), its nonparametric estimates $\widehat{\mu}_M(x^{(i)})$ (points), and the parametric model $\mu(x, \widehat{c}_{N_0,M})$ (thin line)

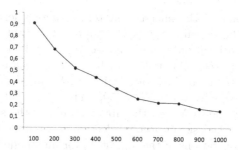

Fig. 3.3 Estimation error $ERR(\widehat{\mu}_M(x))$ depending on the number of measurements M

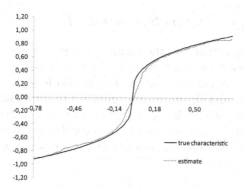

Fig. 3.4 The true characteristic $\mu(x) = \sqrt[3]{x}$ and its nonparametric estimates $\widehat{\mu}_M(x^{(i)})$

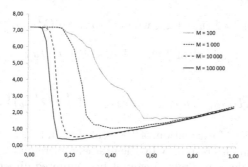

Fig. 3.5 Relationship between the estimation error $ERR(\widehat{\mu}_M(x))$ and the bandwidth parameter h

Conclusions

The nonlinear characteristic of Wiener system is successfully recovered from the input-output data under small amount of a priori information. The estimate (3.26) is consistent under IIR dynamics, non-Gaussian input and non-invertible functions. It is universal in the sense that it can be applied, under quite mild conditions, for Hammerstein systems and for Wiener-Hammerstein systems, as it will be shown in the next Chapter. The strategy allows to decompose of the identification task of block-oriented system and can support estimation of parameters. Computing of both the estimate $\hat{\mu}_M(x)$ and the distance $\delta_k(x)$ has the numerical complexity $O(M)$, and can be performed in recursive or semi-recursive version (see [50]). The main limitation is assumed knowledge of the value of λ, i.e., the upper bound of the speed of impulse response decaying. The issue of proper selection of λ is open for further studies. Potential generalizations of the algorithm for unbounded-input case and for other kernel functions seem to be promising.

3.3.3   Averaged Derivative Method

The interesting new attempt to the impulse response estimation of the linear block in Wiener system was presented in [147]. It is assumed that the input probability density $f(u)$ has compact support, both $\mu()$ and $f()$ have continuous derivatives, and the linear dynamics is FIR with known order s. We emphasize that similarly as in the correlation-based algorithm (see Section 3.2.2) the characteristic $\mu()$ need not to be invertible and moreover the input density is not assumed to be Gaussian. The idea follows from the chain rule. Introducing the vectors

$$\underline{u}_k = (u_k, u_{k-1}, ..., u_{k-s})^T \text{ and } \underline{\lambda} = (\lambda_0, \lambda_1, ..., \lambda_s)^T$$

one can describe the Wiener system by the following formula

$$y_k = F(\underline{u}_k) + z_k, \text{ where } F(\underline{u}_k) = \mu(\underline{\lambda}^T \underline{u}_k).$$

Let $D_F(\underline{u})$ be the gradient of $F()$. It holds that $D_F(\underline{u}_k) = \mu'(\underline{\lambda}^T \underline{u}_k)\underline{\lambda}$ and consequently

$$E\{D_F(\underline{u}_k)\} = c_0\underline{\lambda}, \text{ where } c_0 = E\left\{\mu'(\underline{\lambda}^T \underline{u}_k)\right\}. \tag{3.27}$$

It leads to the idea of estimation of the scaled vector $\underline{\lambda}$, including the true elements of the impulse response, by the gradient averaging. Since for a given $\underline{u}_k$, $\mu'(\underline{\lambda}^T \underline{u}_k)$ is unknown, $D_F(\underline{u}_k)$ cannot be computed directly. Introducing $f_{\underline{u}}()$ – the joint probability density of $\underline{u}_k$, the property (3.27) can be transformed to the more applicable form ([147])

$$E\{y_k D_f(\underline{u}_k)\} = c_1\underline{\lambda}, \text{ where } c_1 = \frac{1}{2}E\left\{f(u_k)\mu'(\underline{\lambda}^T \underline{u}_k)\right\},$$

and $D_f(u)$ is a gradient of $f_{\underline{u}}()$. Since for white u_k we have $f_{\underline{u}}(\underline{u}_k) = \prod_{j=0}^{s} f(u_{k-j})$, it leads to the following scalar estimates of the impulse response

$$\widehat{\lambda}_j = \frac{1}{N} \sum_{k=1}^{N} y_k d_{f,j}(\underline{u}_k), \text{ where } d_{f,j}(u_k) = D_f(\underline{u}_k)[j] = f'(u_{k-j}) \prod_{i=0, i \neq j}^{s} f(u_{k-i}).$$

$$(3.28)$$

If the input probability density function $f(u)$ is unknown, it can be simply estimated, e.g., by the kernel method. The open question is generalization of the approach for IIR linear subsystems and correlated input cases.

3.3.4 Mixed Approach

In [108] the Wiener system with a finite memory (of length p) is identified by a two-step parametric-nonparametric approach. The method permits recovery of a wide-class of nonlinearities which need not be invertible. Furthermore, the algorithm allows non-Gaussian input signals and the presence of additive output measurement noise. We begin with splitting the measurement set into two disjoint subsets. Let I_1, I_2 denote sets of measurements indexes in these subsets (n_1, n_2 are the numbers of elements in I_1 and I_2, respectively). The impulse response function of the linear part is recovered via minimization ($\hat{\lambda} = \arg\min_\lambda \hat{Q}(\lambda)$) of the following nonlinear least squares criterion

$$\hat{Q}(\lambda) = \frac{1}{n_2} \sum_{i \in I_2} \{y_i - \hat{\mu}(v_i(\lambda); \lambda)\}^2, \qquad (3.29)$$

where the system non-linearity $\mu(\cdot)$ is pre-estimated by a pilot non-parametric kernel regression estimate of the form

$$\hat{\mu}(v; \lambda) = \frac{\sum_{j \in I_1} y_j K(\frac{v - v_j(\lambda)}{h})}{\sum_{j \in I_1} K(\frac{v - v_j(\lambda)}{h})}. \qquad (3.30)$$

and where $h = h(n_1)$ is the bandwidth parameter. The estimate $\hat{\lambda}$ is then used to recover interconnection v_n by the estimate $v_n(\hat{\lambda})$. In consequence, using the sequence of pairs $(v_n(\hat{\lambda}), y_n)$, the non-parametric kernel estimate of the nonlinearity $\mu(\cdot)$ is built in a following way

$$\hat{\mu}(v) = \frac{\sum_{j \in I_1} y_j K(\frac{v - v_j(\hat{\lambda})}{h})}{\sum_{j \in I_1} K(\frac{v - v_j(\hat{\lambda})}{h})}.$$

The consistency of the estimates $\hat{\lambda}$ and $\hat{\mu}(v)$ is established in [108] for a large class of input signals and non-invertible nonlinearities.

3.4   Final Remarks

The principal question in Wiener system identification problem is selection
of adequate method. The scope of application of each estimates is limited by
specific set of associated assumptions. Most of them requires a priori known
parametric type of model, Gaussian input, FIR dynamics or invertible characteristic. In fact, the authors address particular cases, and the problems they
solve are quite different (see references below). Since the general Wiener system identification problem includes many difficult aspects, existence of one
universal algorithm cannot be expected. In the light of this, the nonparametric approach seems to be good tool, which allows for combining selected
parametric and nonparametric methods, depending on specificity of the particular task. Moreover, pure nonparametric estimates are the only possible
choice, when the prior knowledge of the system is poor.

Chapter 4
Wiener-Hammerstein (Sandwich) System

4.1 Introduction

In this chapter we address the problem of nonlinearity recovering in the system of Wiener-Hammerstein structure (see Fig. 4.1). It consists of one static nonlinear block with the characteristic $\mu()$, surrounded by two linear dynamic components with the impulse responses $\{\lambda_j\}_{j=0}^{\infty}$ and $\{\gamma_j\}_{j=0}^{\infty}$, respectively. Such a structure, and its particular cases (Wiener systems and Hammerstein systems), are widely considered in the literature because of numerous potential applications in various domains of science and technology (see e.g. [35]). The Wiener and Wiener-Hammerstein models allow for a good approximation of many real processes ([18], [74], [141], [146], [151], [152]). It was noticed that the nonparametric algorithms proposed in Section 3.2.3 (see ([90]), ([97]), ([93])) and in ([44])) for a Wiener system, can be adopted, after slight modification, for a broad class of Wiener-Hammerstein (sandwich) systems. All the assumptions taken therein remain the same. The algorithms work under poor prior knowledge of subsystems and excitations and in contrast to earlier literature items concerning sandwich and Wiener system identification:

- the input sequence need not to be a Gaussian white noise,
- the nonlinear characteristic is not assumed to be invertible,
- the IIR linear dynamic blocks are admitted,
- the algorithm is of nonparametric nature (see e.g. [50]), i.e. it is not assumed that the subsystems can be described with the use of finite and known number of parameters. In consequence, the estimates are free of the possible approximation error, or this error can be made arbitrarily small by proper selection of tuning parameters.

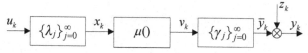

Fig. 4.1 Wiener-Hammerstein (sandwich) system

G. Mzyk, *Combined Parametric-Nonparametric Identification*
of Block-Oriented Systems, Lecture Notes in Control and Information Sciences 454,
DOI: 10.1007/978-3-319-03596-3_4, © Springer International Publishing Switzerland 2014

Firstly, in Section 4.2, we show intuitively, why the censored algorithm (3.14) proposed for Wiener system can be successfully applied for Hammerstein systems and Wiener-Hammerstein (sandwich) systems. In Section 4.3, the problem is formulated in detail and the assumptions imposed on signals and system components are discussed. Then, in Section 4.4 we present two nonparametric kernel-based estimates of the nonlinearity in Wiener-Hammerstein system, and analyze their properties. Finally, in Section 4.5, we illustrate their behaviour in simulation example, for various numbers of observations and values of tuning parameters.

4.2 Preliminaries

In this section we explain that under Assumption 3.6 concerning static characteristic, the estimate (3.14) can be also applied for Hammerstein systems and for Wiener-Hammerstein systems.

Hammerstein System

For the Hammerstein system (Fig. 4.2) described by

$$y_k = \sum_{j=0}^{\infty} \gamma_j \mu(x_{k-j}) + z_k, \qquad (4.1)$$

we assume that the unknown impulse response $\{\gamma_j\}_{j=0}^{\infty}$ fulfils conditions analogous to Assumptions 3.2, and 3.5, i.e., $|\gamma_j| \leqslant c_1 \lambda^j$, and $G = \sum_{j=0}^{\infty} \gamma_j = 1$. For Lipschitz function $\mu()$ we simply get

$$|\bar{y}_k - \mu(x)| = \left| \sum_{j=0}^{\infty} \gamma_j \mu(x_{k-j}) - \sum_{j=0}^{\infty} \gamma_j \mu(x) \right| =$$

$$= \left| \sum_{j=0}^{\infty} \gamma_j (\mu(x_{k-j}) - \mu(x)) \right| =$$

$$= \left| \sum_{j=0}^{k-1} \gamma_j (\mu(x_{k-j}) - \mu(x)) + \sum_{j=k}^{\infty} \gamma_j (\mu(x_{k-j}) - \mu(x)) \right| \leqslant$$

$$\leqslant \sum_{j=0}^{k-1} |\gamma_j| |\mu(x_{k-j}) - \mu(x)| + 2u_{\max} l \sum_{j=k}^{\infty} |\lambda_j| \leqslant$$

$$\leqslant l\delta_k(x) + \frac{l\lambda^k}{1-\lambda} = l\Delta_k(x), \qquad (4.2)$$

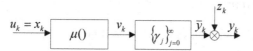

Fig. 4.2 Hammerstein system

which means that for a given x, the selection condition (3.10) is fulfilled and the noise-free output $\bar{y}_k$ is located close to $\mu(x)$. In the next section we will show that under (6.4), the convergence (3.16) given in Theorem 3.3 can be proved for Hammerstein system, with the use of the same technique.

Wiener-Hammerstein System

Now, let us consider a tandem three-element connection shown in Fig. 4.1, where u_k and y_k is a measurable system input and output at time k respectively, z_k is a random noise, $\mu()$ is the unknown characteristic of the static nonlinearity and $\{\lambda_j\}_{j=0}^{\infty}$, $\{\gamma_j\}_{j=0}^{\infty}$ – the unknown impulse responses of the two linear dynamic components. By assumption, the interaction signals x_k and v_k are not available for measurements. The system is described as follows

$$y_k = \sum_{j=0}^{\infty} \gamma_j v_{k-j} + z_k, \quad v_k = \mu\left(\sum_{j=0}^{\infty} \lambda_j u_{k-j}\right). \qquad (4.3)$$

Similarly as for Hammerstein system, under Assumption 3.6, we get

$$|\bar{y}_k - \mu(x)| = \left|\sum_{i=0}^{\infty} \gamma_i \mu(x_{k-i}) - \sum_{i=0}^{\infty} \gamma_i \mu(x)\right| =$$

$$= \left|\sum_{i=0}^{\infty} \gamma_i \mu\left(\sum_{j=0}^{\infty} \lambda_j u_{k-i-j}\right) - \sum_{i=0}^{\infty} \gamma_i \mu\left(\sum_{j=0}^{\infty} \lambda_j x\right)\right| =$$

$$= \left|\sum_{i=0}^{\infty} \gamma_i \left[\mu\left(\sum_{j=0}^{\infty} \lambda_j u_{k-i-j}\right) - \mu\left(\sum_{j=0}^{\infty} \lambda_j x\right)\right]\right| \leqslant$$

$$\leqslant l \sum_{i=0}^{\infty} |\gamma_i| \left|\sum_{j=0}^{\infty} \lambda_j (u_{k-i-j} - x)\right| \leqslant$$

$$\leqslant l \sum_{i=0}^{\infty} |\gamma_i| \sum_{j=0}^{\infty} |\lambda_j| |u_{k-i-j} - x| = l \sum_{i=0}^{\infty} \varkappa_i |u_{k-i} - x|,$$

where the sequence $\{\varkappa_i\}_{i=0}^{\infty}$ is the convolution of $\{|\gamma_i|\}_{i=0}^{\infty}$ with $\{|\lambda_j|\}_{i=0}^{\infty}$, which obviously fulfills the condition $|\varkappa_i| \leqslant \lambda^i$.

4.3   Assumptions

For a Wiener-Hammerstein system we assume that:

Assumption 4.1. *The input $\{u_k\}$ is an i.i.d., bounded ($|u_k| < u_{max}$; unknown $u_{max} < \infty$) random process, and there exists a probability density of the input, say $\vartheta_u(u_k)$, which is a continuous and strictly positive function around the estimation point x, i.e., $\vartheta_u(x) \geqslant \varepsilon > 0$.*

Assumption 4.2. *The unknown impulse responses $\{\lambda_j\}_{j=0}^{\infty}$ and $\{\gamma_j\}_{j=0}^{\infty}$ of the linear IIR filters are both exponentially upper bounded, that is*

$$|\lambda_j| \leqslant c_1 \lambda^j, \ |\gamma_j| \leqslant c_1 \lambda^j, \ some \ unknown \ 0 < c_1 < \infty, \qquad (4.4)$$

where $0 < \lambda < 1$ is an a priori known constant.

Assumption 4.3. *The nonlinear characteristic $\mu(x)$ is a Lipschitz function, i.e., it exists a positive constant $l < \infty$, such that for each $x_a, x_b \in R$ it holds that*

$$|\mu(x_a) - \mu(x_b)| \leqslant l \, |x_a - x_b| \,.$$

Assumption 4.4. *The output noise $\{z_k\}$ is a zero-mean stationary and ergodic process, which is independent of the input $\{u_k\}$.*

Assumption 4.5. *For simplicity of presentation we also let $L \triangleq \sum_{j=0}^{\infty} \lambda_j = 1$, $G \triangleq \sum_{j=0}^{\infty} \gamma_j = 1$, and $u_{max} = \frac{1}{2}$.*

The goal is to estimate the unknown characteristic of the nonlinearity $\mu(x)$ on the interval $x \in (-u_{max}, u_{max})$ on the basis of N input-output measurements $\{(u_k, y_k)\}_{k=1}^{N}$ of the whole Wiener-Hammerstein system.

Similarly as for Wiener system from Asssumptions 4.1 and 4.2 it holds that $|x_k| < x_{max} < \infty$, where $x_{max} \triangleq u_{max} \sum_{j=0}^{\infty} |\lambda_j|$. The condition (4.4), with unknown c_1, is rather not restrictive, and characterizes the class of stable objects. In particular case of FIR linear dynamic blocks, Assumption 4.2 is fulfilled for arbitrarily small $\lambda > 0$.

As regards the Assumption 4.5, we note, that the class of Wiener-Hammerstein systems composed by series connection of linear filters with the impulse responses $\{\overline{\lambda}_j\} = \{\frac{\lambda_j}{\beta}\}_{j=0}^{\infty}$, $\{\overline{\gamma}_j\} = \{\frac{\gamma_j}{\alpha}\}_{j=0}^{\infty}$ and the nonlinearities $\overline{\mu}(x) = \alpha\mu(\beta x)$ are, for $\alpha, \beta \neq 0$, indistinguishable from the input-output point of view.

Remark 4.1. *If the technical Assumption 4.5 is not fulfilled, i.e., the gains $L = \sum_{j=0}^{\infty} \lambda_j$ or $G = \sum_{j=0}^{\infty} \gamma_j$ are not unit, then only the scaled and dilated version $G\mu(Lx)$ of the true system characteristic $\mu(x)$ can be identified. The constants G and L are not identifiable, since the internal signals x_k and v_k cannot be measured.*

4.4   The Algorithms

For a Wiener-Hammerstein system we apply and compare the following two nonparametric kernel-based estimates of the nonlinear characteristic $\mu()$, proposed before for a Wiener system in ([90]) and ([44]), respectively.

$$\widehat{\mu}_N^{(1)}(x) = \frac{\sum_{k=1}^{N} y_k \cdot K\left(\frac{\sum_{j=0}^{k}|u_{k-j}-x|\lambda^j}{h(N)}\right)}{\sum_{k=1}^{N} K\left(\frac{\sum_{j=0}^{k}|u_{k-j}-x|\lambda^j}{h(N)}\right)}, \tag{4.5}$$

$$\widehat{\mu}_N^{(2)}(x) = \frac{\sum_{k=1}^{N} y_k \prod_{i=0}^{p} K\left(\frac{x-u_{k-i}}{h(N)}\right)}{\sum_{k=1}^{N} \prod_{i=0}^{p} K\left(\frac{x-u_{k-i}}{h(N)}\right)}. \tag{4.6}$$

In (4.5) and (4.6) $K()$ is a bounded kernel function with compact support, i.e., it fulfills the following conditions

$$\int_{-\infty}^{\infty} K(x)dx = 1,$$
$$\sup_{x} |K(x)| < \infty, \tag{4.7}$$
$$K(x) = 0 \text{ for } |x| > x_0, \text{ some } x_0 < \infty.$$

The sequence $h(N)$ (bandwidth parameter) is such that

$$h(N) \to 0, \text{ as } N \to \infty.$$

The following theorem holds.

Theorem 4.1. *If $h(N) = d(N)\log_\lambda d(N)$, where $d(N) = N^{-\gamma(N)}$, and $\gamma(N) = \left(\log_{1/\lambda} N\right)^{-w}$, then for each $w \in \left(\frac{1}{2}, 1\right)$ the estimate (4.5) is consistent in the mean square sense, i.e., it holds that*

$$\lim_{N\to\infty} E\left(\widehat{\mu}_N^{(1)}(x) - \mu(x)\right)^2 = 0. \tag{4.8}$$

Proof. For the proof see Appendix A.17.   ∎

In contrast to $\widehat{\mu}_N^{(1)}(x)$, the estimate $\widehat{\mu}_N^{(2)}(x)$ uses the FIR(p) approximation of the linear subsystems. We will show that since the linear blocks are asymptotically stable, the approximation of $\mu()$ can be made with arbitrary accuracy, i.e., by selecting p large enough. Let us introduce the following regression-based approximation of the true characteristic $\mu()$

$$m_p(x) = E\{y_k|u_k = u_{k-1} = ... = u_{k-2p+1} = x\} \tag{4.9}$$

and the constants

$$g_p = \sum_{i=0}^{p-1} \gamma_i, \qquad l_p = \sum_{j=0}^{p-1} \lambda_j.$$

The following theorem holds.

Theorem 4.2. *If $K()$ satisfy (4.7) then it holds that*

$$\widehat{\mu}_N^{(2)}(x) \to m_p(l_p x) \text{ in probability,} \tag{4.10}$$

as $N \to \infty$, at every point x, for which $\vartheta_u(x) > 0$ provided that

$$N h^{2p}(N) \to \infty, \text{ as } N \to \infty.$$

Proof. *The proof is a consequence of (A.35) and the proof of Theorem 1 in [44].*   ∎

From (A.39) we obtain that

$$m_p(x) =$$

$$= E\left\{ \sum_{i=0}^{p-1} \gamma_i \mu(x_{k-i}) + \varsigma \,\middle|\, u_k = \ldots = u_{k-2p+1} = x \right\}$$

where $\varsigma = \sum_{i=p}^{\infty} \gamma_i \mu(x_{k-i})$. Moreover, since $x_k = \sum_{j=0}^{p-1} \lambda_j u_{k-j} + \xi$, where $\xi = \sum_{j=p}^{\infty} \lambda_j u_{k-j}$ it holds that

$$|m_p(l_p x) - \mu(l_p x)| =$$
$$|E\{g_p \mu(l_p x + \xi) + \varsigma\} - \mu(l_p x)| \le$$
$$\le E |\{g_p \mu(l_p x + \xi) + \varsigma\} - \mu(l_p x)| \le$$
$$\le |g_p - 1| \left(l E u_k + E \mu(x_k) \right),$$

and under stability of linear components (see Assumptions 4.2 and 4.5) we have

$$|g_p - 1| \le c_0^p, \text{ some } |c_0| < 1.$$

Consequently,

$$\widehat{\mu}_N^{(2)}(x) \to \mu(l_p x) + \epsilon_p$$

in probability, as $N \to \infty$, where $\epsilon_p = c_0^p \left(l u_{\max} + v_{\max} \right) \phi(x)$, and $|\phi(x)| \le 1$. Since $\lim_{p\to\infty} l_p = 1$, and $\lim_{p\to\infty} \epsilon_p = 0$ we conclude that (4.10) is constructive in the sense that the approximation model of $\mu()$ can have arbitrary accuracy by proper selection of p.

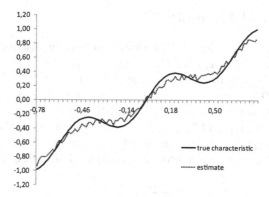

Fig. 4.3 The true characteristic $\mu(x) = x + 0.2\sin(10x)$ and its nonparametric estimate $\widehat{\mu}_N^{(1)}(x)$

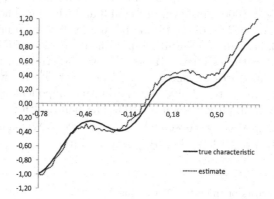

Fig. 4.4 The true characteristic $\mu(x) = x + 0.2\sin(10x)$ and its nonparametric estimate $\widehat{\mu}_N^{(2)}(x)$

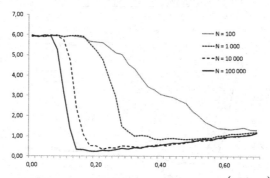

Fig. 4.5 Relationship between the estimation error $ERR\left(\widehat{\mu}_N^{(1)}(x)\right)$ and the bandwidth parameter h

4.5   Numerical Example

In the computer experiment we generated uniformly distributed i.i.d. input sequence $u_k \sim U[-1,1]$ and the output noise $z_k \sim U[-0.1,0.1]$. We simulated the IIR linear dynamic subsystems $x_k = 0.5x_{k-1} + 0.5u_k$ and $\bar{y}_k = 0.5\bar{y}_{k-1} + 0.5v_k$, i.e. $\lambda_j = \gamma_j = 0.5^{j+1}$, $j = 0,1,...,\infty$, sandwiched with the not invertible static nonlinear characteristic $\mu(x) = x + 0.2\sin(10x)$. The nonparametric estimates (4.5) and (4.6) were computed on the same simulated data $\{(u_k, y_k)\}_{k=1}^{N}$. In Assumption 4.2 we assumed $\lambda = 0.8$ and in (4.6) we took $p = 3$. The estimation error was computed according the rule

$$ERR\left(\widehat{\mu}_N(x)\right) = \sum_{i=1}^{N_0} \left(\widehat{\mu}_N(x^{(i)}) - \mu(x^{(i)})\right)^2, \qquad (4.11)$$

where $\{x^{(i)}\}_{i=1}^{N_0}$ is the grid of equidistant estimation points. The results of estimation for $N = 1000$ are shown in Fig. 4.3 and Fig. 4.4. The routine was repeated for various values of the tuning parameter h. As can be seen in Fig. 4.5 and Fig. 4.6, according to intuition, improper selection of h increases the variance or bias of the estimate. Table 1 shows the errors (4.11) of $\widehat{\mu}_N^1(x)$ and $\widehat{\mu}_N^2(x)$. It illustrates advantages of $\widehat{\mu}_N^1(x)$ over $\widehat{\mu}_N^2(x)$, when number of measurements tends to infinity and the linear component in the Wiener system has infinite impulse response (IIR). The bandwidth parameters was set according to $h(N) = N^{-\left(\log_{1/\lambda} N\right)^{-w}} \log_\lambda N^{-\left(\log_{1/\lambda} N\right)^{-w}}$ with $w = 0.75$ in (4.5), and $h(N) = N^{-1/(2p+1)}$ with $p = 5$ in (4.6).

Table 4.1 The errors of the estimates versus N

N	10^2	10^3	10^4	10^5	10^6
$ERR\left(\widehat{\mu}_N^1(x)\right)$	6.1	4.9	0.8	0.5	0.3
$ERR\left(\widehat{\mu}_N^2(x)\right)$	9.8	8.1	4.4	1.1	0.8

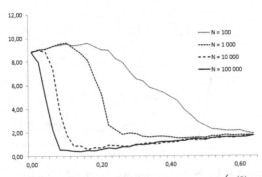

Fig. 4.6 Relationship between the estimation error $ERR\left(\widehat{\mu}_N^{(2)}(x)\right)$ and the bandwidth parameter h

4.6 Final Remarks

The nonlinear characteristic of Wiener-Hammerstein system is successfully recovered from the input-output data under small amount of a priori information. The estimates work under IIR dynamic blocks, non-Gaussian input and for non-invertible characteristics. Since the Hammerstein systems and the Wiener systems are special cases of the sandwich system, the proposed approach is universal in the sense that it can be applied without the prior knowledge of the cascade system structure.

As regards the limit properties, the estimates $\widehat{\mu}_N^{(1)}(x)$ and $\widehat{\mu}_N^{(2)}(x)$ are not equivalent. First of them has slower rate of convergence (logarithmic), but it converges to the true system characteristic, since the model becomes more complex as the number of observations tends to infinity. The main limitation is assumed knowledge of λ, i.e., the upper bound of the impulse response. On the other hand, the rate of convergence of the estimate $\widehat{\mu}_N^{(2)}(x)$ is faster (exponential), but the estimate is biased, even asymptotically. However, its bias can be made arbitrarily small by selecting the cut-off parameter p large enough.

As it was shown in [65], the nonparametric methods allow for decomposition of the identification task of block-oriented system and can support estimation of its parameters. Computing of both estimates $\widehat{\mu}_N^{(1)}(x)$, $\widehat{\mu}_N^{(2)}(x)$ and the distance $\delta_k(x)$ has the numerical complexity $O(N)$, and can be performed in recursive or semi-recursive version (see [50]).

Chapter 5
Additive NARMAX System

5.1 Statement of the Problem

5.1.1 The System

In this chapter we consider a scalar, discrete-time, asymptotically stable non-linear dynamic system shown in Fig. 5.1, and described by the following equation (cf. [2], [21], [36], [87], [88], [89], [99]):

$$y_k = \sum_{j=1}^{p} \lambda_j \eta(y_{k-j}) + \sum_{i=0}^{n} \gamma_i \mu(u_{k-i}) + z_k, \qquad (5.1)$$

where

$$\mu(u) = \sum_{t=1}^{m} c_t f_t(u), \qquad (5.2)$$

$$\eta(y) = \sum_{l=1}^{q} d_l g_l(y).$$

The structure is well known in the literature as the additive NARMAX model ([21]). The signals y_k, u_k and z_k are the output, the input and the noise, respectively. The system in Fig. 5.1 is more general then often met in the literature Hammerstein system. The Hammerstein system is its special case, when the function $\eta()$ is linear (see the Appendix 8.1). The additive NAR-MAX system is also not equivalent to widely considered in the literature Wiener-Hammerstein (sandwich) system, where two linear dynamic blocks surround one static nonlinearity. In spite of many potential possibilities of applications in various domains ([54], [2], [158], [135], [118], [85]), relatively small attention has been paid to this structure in the literature. We take the following assumptions.

G. Mzyk, *Combined Parametric-Nonparametric Identification*
of Block-Oriented Systems, Lecture Notes in Control and Information Sciences 454,
DOI: 10.1007/978-3-319-03596-3_5, © Springer International Publishing Switzerland 2014

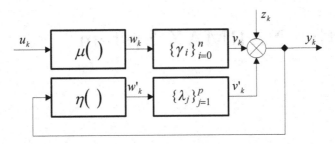

Fig. 5.1 The additive NARMAX system

Assumption 5.1. *The input process $\{u_k\}$ is a sequence of i.i.d. bounded random variables, i.e., it exists (unknown) $u_{\max}$, such that $|u_k| < u_{\max} < \infty$.*

Assumption 5.2. *The system as a whole is asymptotically stable, i.e. the system output y_k is bounded; $|y_k| < y_{\max} < \infty$.*

Assumption 5.3. *The static nonlinear characteristics are of a given parametric form*

$$\mu(u) = \sum_{t=1}^{m} c_t f_t(u) \tag{5.3}$$

$$\eta(y) = \sum_{l=1}^{q} d_l g_l(y)$$

where $f_1(),...,f_m()$ and $g_1(),...,g_q()$ are a priori known linearly independent basis functions, bounded on the operating range, i.e.

$$|f_t(u)| \le p_{\max}, \tag{5.4}$$

$$|g_l(y)| \le p_{\max},$$

for $|u| \le u_{\max}$ and $|y| \le y_{\max}$, some constant $p_{\max}$.

Assumption 5.4. *The linear dynamic blocks have finite impulse responses, i.e.,*

$$v_k = \sum_{i=0}^{n} \gamma_i w_{k-i} \tag{5.5}$$

$$v'_k = \sum_{j=1}^{p} \lambda_j w'_{k-j}$$

with known orders n and p.

Assumption 5.5. *The output noise $\{z_k\}$ is correlated linear process. It can be written as*

$$z_k = \sum_{i=0}^{\infty} \omega_i \varepsilon_{k-i}, \tag{5.6}$$

where $\{\varepsilon_k\}$ is some unknown zero-mean $(E\varepsilon_k = 0)$ and bounded $(|\varepsilon_k| < \varepsilon_{\max} < \infty)$ i.i.d. process, independent of the input $\{u_k\}$, and $\{\omega_i\}_{i=0}^{\infty}$ $(\sum_{i=0}^{\infty} |\omega_i| < \infty)$ is the unknown stable linear filter.

Assumption 5.6. *Only the input $\{u_k\}$ and the output of the whole system $\{y_k\}$ are accessible for measurements.*

Let

$$\Lambda = (\lambda_1, .., \lambda_p)^T, \tag{5.7}$$
$$\Gamma = (\gamma_0, ..., \gamma_n)^T,$$
$$c = (c_1, ..., c_m)^T,$$
$$d = (d_1, ..., d_q)^T,$$

denote true (unknown) parameters of the system. Obviously, the input-output description of the system, given by (5.1)-(5.2) is not unique. For each pair of constants $\bar{\alpha}$ and $\bar{\beta}$, the systems with parameters Λ, Γ, c, d and $\bar{\beta}\Lambda$, $\bar{\alpha}\Gamma$, $c/\bar{\alpha}$, $d/\bar{\beta}$ cannot be distinguished, i.e., are equivalent (see (5.1)-(5.2)). For the uniqueness of the solution we introduce the following technical assumption (see [2]):
(a) the matrices $\Theta_{\Lambda d} = \Lambda d^T$ and $\Theta_{\Gamma c} = \Gamma c^T$ are not both zero;
(b) $||\Lambda||_2 = 1$ and $||\Gamma||_2 = 1$, where $||.||_2$ is Euclidean vector norm;
(c) first non-zero elements of Λ and Γ are positive.
   Let

$$\theta = (\gamma_0 c_1, ..., \gamma_0 c_m, ..., \gamma_n c_1, ..., \gamma_n c_m, \lambda_1 d_1, ..., \lambda_1 d_q, ..., \lambda_p d_1, ..., \lambda_p d_q)^T$$
$$= (\theta_1, ..., \theta_{(n+1)m}, \theta_{(n+1)m+1}, ..., \theta_{(n+1)m+pq})^T \tag{5.8}$$

be the vector of aggregated parameters (5.1) obtained by inserting (5.2) to (5.1), and let ϕ_k be respective generalized input vector

$$\phi_k = (f_1(u_k), ..., f_m(u_k), ..., f_1(u_{k-n}), ..., f_m(u_{k-n}),$$
$$g_1(y_{k-1}), ..., g_q(y_{k-1}), ..., g_1(y_{k-p}), ..., g_q(y_{k-p}))^T. \tag{5.9}$$

Thanks to above notation, description (5.1)-(5.2) can be simplified to the form $y_k = \phi_k^T \theta + z_k$, which means that the system remains linear with respect to aggregated parameters. For $k = 1, ..., N$ we obtain

$$Y_N = \Phi_N \theta + Z_N \tag{5.10}$$

where $Y_N = (y_1, ..., y_N)^T$, $\Phi_N = (\phi_1, ..., \phi_N)^T$, and $Z_N = (z_1, ..., z_N)^T$.
   The purpose of identification is to recover parameters in Λ, Γ, c and d (given by (5.7)), using the input-output measurements (u_k, y_k) $(k = 1, ..., N)$ of the whole system.

5.1.2   Comments to Assumptions

The representation (5.1) belongs to the class of so-called "equation-error" models, while in practical situations a more complicated case of "output-error" models is often met, i.e.

$$\begin{cases} \overline{y}_k = \sum_{j=1}^{p} \lambda_j \eta(\overline{y}_{k-j}) + \sum_{i=0}^{n} \gamma_i \mu(u_{k-i}), \\ y_k = \overline{y}_k + \delta_k, \end{cases}$$

with zero-mean disturbance δ_k. Since the resulting noise z_k in (5.1) results from nonlinear filtering of δ_k, it can be of relatively high order and can have non-zero mean. First problem was solved in the book only for the restricted class of systems which fulfill additional condition imposed on the elements in the feedback (see Assumption 5.7 in Section 5.5). The second problem can be simply solved when the constant function is appended to the basis $f_1(),...,f_m()$.

To simplify presentation it was assumed that both the input process, the nonlinear characteristics and the noise are bounded. In fact, since further analysis assumes only finite fourth order moments of all signals, the approach can be generalized for Lipschitz nonlinearities and most of popular finite-variance distributions of excitations.

As regards the i.i.d. restriction imposed on the input process, it can be weakened, for invertible processes, by e.g. data pre-filtering and the use of specially designed instrumental variables in parameter identification (see [98]).

5.1.3   Organization of the Chapter

In the Section 5.2, the least squares based identification algorithm (see [2]) will be presented for white disturbances. Then, the reason of its asymptotic bias will be shown for correlated noise. Next, in Section 5.3, asymptotically unbiased, instrumental variables based estimate is proposed. The idea originates from the linear system theory (see e.g. [127] and [154]), where the instrumental variables technique is used for identification of simple one-element linear dynamic objects. The proposed method is then compared with the least squares. In particular, the consistency of the proposed estimate is shown even for correlated disturbances. The form of the optimal instrumental variables is established and the nonparametric method of their approximate generation is described. Also, the asymptotic rate of convergence of the estimate is analyzed.

5.2   Least Squares and SVD Approach

For comparison purposes with the proposed further instrumental variables method we start from presentation of the two-stage algorithm based on least squares estimation of the aggregated parameter vector and decomposition of the obtained result with the use of SVD algorithm (see [2], and [78]). The algorithm is as follows.

Stage 1. Compute the LS estimate

$$\widehat{\theta}_N^{(LS)} = (\Phi_N^T \Phi_N)^{-1} \Phi_N^T Y_N \tag{5.11}$$

of the aggregated parameter vector θ (see (5.8) and (5.10)), and next construct (by the plug-in method) evaluations $\widehat{\Theta}_{\Lambda d}^{(LS)}$ and $\widehat{\Theta}_{\Gamma c}^{(LS)}$ of the matrices $\Theta_{\Lambda d} = \Lambda d^T$ and $\Theta_{\Gamma c} = \Gamma c^T$, respectively (see condition (a) above).

Stage 2. Perform the SVD (Singular Value Decomposition – see the Appendix B.1) of the matrices $\widehat{\Theta}_{\Lambda d}^{(LS)}$ and $\widehat{\Theta}_{\Gamma c}^{(LS)}$:

$$\widehat{\Theta}_{\Lambda d}^{(LS)} = \sum_{i=1}^{\min(p,q)} \delta_i \widehat{\xi}_i \widehat{\zeta}_i^T \tag{5.12}$$

$$\widehat{\Theta}_{\Gamma c}^{(LS)} = \sum_{i=1}^{\min(n,m)} \sigma_i \widehat{\mu}_i \widehat{\nu}_i^T$$

and next compute the estimates of parameters of particular blocks (see (5.7))

$$\widehat{\Lambda}_N^{(LS)} = sgn(\widehat{\xi}_1[\kappa_{\xi_1}])\widehat{\xi}_1, \tag{5.13}$$

$$\widehat{\Gamma}_N^{(LS)} = sgn(\widehat{\mu}_1[\kappa_{\mu_1}])\widehat{\mu}_1,$$

$$\widehat{c}_N^{(LS)} = sgn(\widehat{\mu}_1[\kappa_{\mu_1}])\sigma_1\widehat{\nu}_1,$$

$$\widehat{d}_N^{(LS)} = sgn(\widehat{\xi}_1[\kappa_{\xi_1}])\delta_1\widehat{\zeta}_1,$$

where $x[k]$ denotes k-th element of the vector x and $\kappa_\mathbf{x} = \min\{k : x[k] \neq 0\}$.

The form of SVD representations of the theoretical matrices $\Theta_{\Gamma c} = \Gamma c^T$ and $\Theta_{\Lambda d} = \Lambda d^T$ has the fundamental meaning for the algorithm. Each matrix being the product of two vectors has the rank equal 1, and only one singular value is not zero, i.e.,

$$\Theta_{\Gamma c} = \sum_{i=1}^{\min(n,m)} \sigma_i \mu_i \nu_i^T,$$

and

$$\sigma_1 \neq 0, \quad \sigma_2 = ... = \sigma_{\min(n,m)} = 0,$$

thus

$$\Theta_{\Gamma c} = \sigma_1 \mu_1 \nu_1^T, \tag{5.14}$$

where $\|\mu_1\|_2 = \|\nu_1\|_2 = 1$. Representation of $\Theta_{\Gamma c}$ given by (5.14) is obviously unique. To obtain Γ, which fulfills the condition (b) one can take $\Gamma = \mu_1$, or $\Gamma = -\mu_1$. The condition (c) guarantees uniqueness of Γ. The remaining part of decomposition allows to compute c. The vectors Λ and d can be obtained from $\Theta_{\Lambda d}$ in a similar way.

The Singular Value Decomposition allows to split the aggregated matrices of parameters $\widehat{\Theta}_{\Gamma c}^{(LS)}$ and $\widehat{\Theta}_{\Lambda d}^{(LS)}$ into products of two vectors (see (5.12)) and estimating $\widehat{\Gamma}_N^{(LS)}\widehat{c}_N^{(LS)T}$ and $\widehat{\Lambda}_N^{(LS)}\widehat{d}_N^{(LS)T}$ according to (5.13). It was shown in [2] that

$$(\widehat{\mu}_1, \sigma_1\widehat{\nu}_1) = \arg \min_{c\in R^m, \Gamma\in R^n} \left\|\widehat{\Theta}_{\Gamma c}^{(LS)} - \Gamma c^T\right\|^2, \qquad (5.15)$$

and for the noise-free case ($z_k \equiv 0$) the estimates (5.13) are equal to the true system parameters, i.e.,

$$\widehat{\Lambda}_N^{(LS)} = \Lambda, \qquad (5.16)$$
$$\widehat{\Gamma}_N^{(LS)} = \Gamma,$$
$$\widehat{c}_N^{(LS)} = c,$$
$$\widehat{d}_N^{(LS)} = d.$$

Moreover, if the noise $\{z_k\}$ is an i.i.d. process, independent of the input $\{u_k\}$, then it holds that

$$\widehat{\Lambda}_N^{(LS)} \to \Lambda, \qquad (5.17)$$
$$\widehat{\Gamma}_N^{(LS)} \to \Gamma,$$
$$\widehat{c}_N^{(LS)} \to c,$$
$$\widehat{d}_N^{(LS)} \to d,$$

with probability 1, as $N \to \infty$.

Remark 5.1. *For a less sophisticated linear ARMAX model*

$$y_k = \sum_{j=1}^{p} d\lambda_j y_{k-j} + \sum_{i=0}^{n} c\gamma_i u_{k-i} + z_k,$$

where c and d are scalar constants, the vector (5.8) reduces to $\theta = \left(\Theta_{\Gamma c}^T, \Theta_{\Lambda d}^T\right)^T$, with the single column matrices $\Theta_{\Gamma c}$ and $\Theta_{\Lambda d}$. Consequently, the estimate (5.11) plays the role of standard least squares method and the SVD decomposition in (5.12) guarantees normalization, i.e. $\|\Lambda\|_2 = 1$ and $\|\Gamma\|_2 = 1$.

By taking (5.10) and (5.11) into account, the estimation error of the vector θ by the least squares can be expressed as follows

$$\Delta_N^{(LS)} = \widehat{\theta}_N^{(LS)} - \theta = \tag{5.18}$$

$$\left(\Phi_N^T \Phi_N\right)^{-1} \Phi_N^T Z_N$$

$$= \left(\frac{1}{N}\sum_{k=1}^{N}\phi_k\phi_k^T\right)^{-1}\left(\frac{1}{N}\sum_{k=1}^{N}\phi_k z_k\right).$$

If $\{z_k\}$ is a zero-mean white noise with finite variance, independent of $\{u_k\}$, then all elements of the vector Z_N are independent of the elements of the matrix Φ_N and from ergodicity of the noise and the process $\{\phi_k\}$ (see the Appendix B.8) it holds that $\Delta_N^{(LS)} \to 0$ with probability 1, as $N \to \infty$. Nevertheless, if $\{z_k\}$ is correlated, i.e., $Ez_k z_{k+i} \neq 0$ for some $i \neq 0$, then the LS estimate (5.11) of θ is not consistent because of the dependence between z_k, a the values $g_l(y_{k-i})$ $(l = 1, ..., q, i = 1, ..., p)$ included in ϕ_k. Consequently, the estimates given by (5.13) are not consistent, too.

5.3 Instrumental Variables Approach

Let us assume that we have given, or we are able to generate, additional matrix Ψ_N of instrumental variables, which fulfills (even for correlated z_k) the following conditions (see [154], [33], [150], [58], [127], [79], [129], [65]):
(C1): $dim\Psi_N = \dim\Phi_N$, and the elements of $\Psi_N = (\psi_1, \psi_2, ..., \psi_N)^T$, where $\psi_k = (\psi_{k,1}, \psi_{k,2}, ..., \psi_{k,m(n+1)+pq})^T$, are commonly bounded, i.e., there exists $0 < \psi_{\max} < \infty$ such that $|\psi_{k,j}| \leq \psi_{\max}$ $(k = 1...N, j = 1...m(n + 1) + pq)$ and $\psi_{k,j}$ are ergodic, not necessary zero-mean, processes (see the Appendix B.8)
(C2): there exists $\mathrm{Plim}(\frac{1}{N}\Psi_N^T \Phi_N) = E\psi_k\phi_k^T$ and the limit is not singular, i.e., $\det\{E\psi_k\phi_k^T\} \neq 0$
(C3): $\mathrm{Plim}(\frac{1}{N}\Psi_N^T Z_N) = E\psi_k z_k$ and $E\psi_k z_k = cov(\psi_k, z_k) = 0$ (see Assumption 5.5).

Lemma 5.1. *The necessary condition for existence of the instrumental variables matrix Ψ_N, which fulfills (C2) is asymptotic non-singularity of the matrix $\frac{1}{N}\Phi_N^T\Phi_N$, i.e. persistent excitation of the regressors $\{\phi_k\}$.*

Proof. For the proof see the Appendix A.18. ∎

Remark 5.2. *Fulfillment of (C1) and (C3) can be simply guaranteed in practice, if the instruments $\{\psi_k\}$ are generated by linear or nonlinear filtering of the input process $\{u_k\}$, without usage of the noise-corrupted output measurements $\{y_k\}$. Unfortunately, for the considered system, including nonlinear feedback, fulfillment of (C2) is more problematic, and cannot be proved in general case. It is assumed that the necessary condition given by Lemma 5.1 holds, i.e. the least squares based estimate is well defined. Since $\mathrm{Plim}\det(\frac{1}{N}\Phi_N^T\Phi_N) \neq 0$, the ideas of instruments generation presented below are based on the postulate $\mathrm{Plim}(\frac{1}{N}\Psi_N^T\Phi_N) = \mathrm{Plim}(\frac{1}{N}\Phi_N^T\Phi_N)$, i.e. the noise-*

free system outputs are estimated by the nonparametric method, and included to Ψ_N.

After left hand side multiplying of (5.10) by Ψ_N^T we get

$$\Psi_N^T Y_N = \Psi_N^T \Phi_N \theta + \Psi_N^T Z_N.$$

Taking into account conditions (C1)÷(C3) we propose to replace the LS estimate, given by (5.11), and computed in Stage 1 (see Section 5.2) with the instrumental variables estimate

$$\widehat{\theta}_N^{(IV)} = (\Psi_N^T \Phi_N)^{-1} \Psi_N^T Y_N. \tag{5.19}$$

Stage 2 is analogous, i.e., the SVD decomposition is made for the estimates $\widehat{\Theta}_{Ad}^{(IV)}$ and $\widehat{\Theta}_{\Gamma c}^{(IV)}$ of matrices Θ_{Ad} and $\Theta_{\Gamma c}$, obtained on the basis of $\widehat{\theta}_N^{(IV)}$.

5.4 Limit Properties

For the algorithm (5.19) the estimation error of aggregated parameter vector θ has the form

$$\Delta_N^{(IV)} = \widehat{\theta}_N^{(IV)} - \theta = \tag{5.20}$$

$$= (\Psi_N^T \Phi_N)^{-1} \Psi_N^T Z_N = \left(\frac{1}{N} \sum_{k=1}^{N} \psi_k \phi_k^T \right)^{-1} \left(\frac{1}{N} \sum_{k=1}^{N} \psi_k z_k \right).$$

Theorem 5.1. *Under (C1)÷(C3), the estimate (5.19) converges in probability to the true parameters of the system, independently of the autocorrelation of the noise z_k, i.e.,*

$$P \lim_{N \to \infty} \Delta_N^{(IV)} = 0. \tag{5.21}$$

Proof. For the proof see the Appendix A.19.   ■

Theorem 5.2. *The estimation error $\Delta_N^{(IV)}$ converges to zero with the asymptotic rate $O(\frac{1}{\sqrt{N}})$ in probability (see e.g. Definition B.4 in Appendix B.4), for each strategy of instrumental variables generation, which guarantees fulfilment of (C1)÷(C3).*

Proof. For the proof see the Appendix A.20.   ■

5.5 Optimal Instrumental Variables

Theorem 5.2 gives universal guaranteed asymptotic rate of convergence of the estimate (5.19). Nevertheless, for moderate number of measurements, the error depends on particular instruments used in application. In this section, the optimal form of instruments is established for the special case of NARMAX systems, which fulfils the following assumption concerning $\eta()$ and $\{\lambda_j\}_{j=1}^p$.

Assumption 5.7. *The nonlinear characteristic $\eta()$ is a Lipschitz function, i.e.,*

$$\left|\eta(y^{(1)}) - \eta(y^{(2)})\right| \leq r \left|y^{(1)} - y^{(2)}\right|, \tag{5.22}$$

and

$$\eta(0) = 0. \tag{5.23}$$

Moreover, the constant $r > 0$ is such that

$$\alpha = r \sum_{j=1}^{p} |\lambda_j| < 1. \tag{5.24}$$

Remark 5.3. *The inequalities (5.22) and (5.24) constitute sufficient conditions for system stability. Although the above conditions are not necessary for Assumption 5.2, and the class of systems considered in this section is narrowed, it is still more general, than the class of systems with linear feedback (e.g. ARMAX systems or Hammerstein systems). Several applications of such systems can be found in [14], [20], [39] and [140].*

Example 5.1 *For the 1st order Hammerstein system*

$$y_k = \lambda_1 y_{k-1} + \sum_{i=0}^{n} \gamma_i \mu(u_{k-i}) + z_k$$

we have $\eta(y) = y$, and since $r = 1$, we obtain standard restriction $|\lambda_1| \leq 1$.

Example 5.2 *For the square function in the loop, i.e., for the system*

$$y_k = \lambda_1 y_{k-1}^2 + \sum_{i=0}^{n} \gamma_i \mu(u_{k-i}) + z_k$$

we have $r = 2y_{\max}$, and respective class of admissible dynamics $|\lambda_1| \leq \frac{1}{2y_{\max}}$.

Let us consider the following conditional processes (cf. (5.2))

$$G_{l,k} \triangleq E\{g_l(y_k) \mid \{u_i\}_{i=-\infty}^{k}\} \tag{5.25}$$

where $l = 1, 2, ...q$ and denote

$$\xi_l \triangleq g_l(y) - G_l.$$

It holds that

$$g_l(y_k) = G_{l,k} + \xi_{l,k},$$

and the signals
$$\xi_{l,k} = g_l(y_k) - G_{l,k}, \tag{5.26}$$

for $l = 1, 2, \ldots q$, and $k = 1, 2, \ldots, N$, will be interpreted as the 'noises'. The equation (5.1) can be now presented as follows

$$y_k = \sum_{j=1}^{p} \lambda_j \eta(y_{k-j}) + \sum_{i=0}^{n} \gamma_i \mu(u_{k-i}) + z_k = \tag{5.27}$$

$$A_k\left(\{y_{k-j}\}_{j=1}^{p}\right) + B_k\left(\{u_{k-i}\}_{i=1}^{n}\right) + C_k\left(u_k\right) + z_k,$$

where

$$A_k\left(\{y_{k-j}\}_{j=1}^{p}\right) = \sum_{j=1}^{p} \lambda_j \eta(y_{k-j}),$$

$$B_k\left(\{u_{k-i}\}_{i=1}^{n}\right) = \sum_{i=1}^{n} \gamma_i \mu(u_{k-i}),$$

$$C_k\left(u_k\right) = \gamma_0 \mu(u_k).$$

The random variables A_k, B_k and z_k are independent of the input u_k (see Assumptions 5.3-5.6). For a fixed $u_k = u$ we get $C_k\left(u\right) = \gamma_0 \mu(u)$. The expectation in (5.25) has the following interpretation

$$G_{l,k} = E\{g_l(C_k\left(u_k\right) + A_k\left(\{y_{k-j}\}_{j=1}^{p}\right) \tag{5.28}$$

$$+ B_k\left(\{u_{k-i}\}_{i=1}^{n}\right) + z_k) \mid \{u_i\}_{i=-\infty}^{k}\},$$

and cannot be computed explicitly. However, as it will be shown further, the knowledge of functional relation between $G_{l,k}$ and the characteristics $\mu()$, $\eta()$ is not needed. The most significant are the properties below.

Property (P1): The 'disturbances' $\{\xi_{l,k}\}_{k=1}^{N}$ given by (5.26) are independent of the input process $\{u_k\}$ and are all ergodic (see Appendix B.8).

Mutual independence between $\{\xi_{l,k}\}_{k=1}^{N}$ and $\{u_k\}_{k=-\infty}^{\infty}$ is a direct consequence of definition (5.25). On the basis of Assumptions 5.1, 5.5 and 5.2 we conclude, that the output $\{y_k\}_{k=1}^{N}$ of the system is bounded and ergodic. Thanks to Assumption 5.3, concerning the nonlinear characteristics, the processes $\{g_l\left(y_k\right)\}_{k=1}^{N}$ and $\{G_{l,k}\}_{k=1}^{N}$ $(l = 1, 2, \ldots, q)$ are also bounded and ergodic. Consequently, the 'noises' $\{\xi_{l,k}\}_{k=1}^{N}$ $(l = 1, 2, \ldots, q)$, as the sums of ergodic processes, are ergodic too (see (5.26)).

Property (P2): The processes $\{\xi_{l,k}\}$ are zero-mean.

By definition (5.26) of $\xi_{l,k}$ we simply have

$$E\xi_{l,k} = Eg_l(y_k) - EG_{l,k} =$$

$$= E_{\{u\}_{j=-\infty}^k} E\left\{g_l(y_k) \mid \{u\}_{i=-\infty}^k\right\} -$$

$$- E_{\{u\}_{j=-\infty}^k} E\left\{g_l(y_k) \mid \{u\}_{i=-\infty}^k\right\} = 0.$$

Property (P3): If the instrumental variables $\psi_{k,j}$ are generated by nonlinear filtration

$$\psi_{k,j} = H_j(\{u_i\}_{i=-\infty}^k), \tag{5.29}$$

where the transformations $H_j()$ $(j = 1, 2, ..., m(n+1) + pq)$ guarantee the ergodicity of $\{\psi_{k,j}\}$, then all products $\psi_{k_1,j}\xi_{l,k_2}$ $(j = 1, 2, ..., m(n+1) + pq,$ $l = 1, 2, ..., q)$ are zero-mean, i.e., $E\psi_{k_1,j}\xi_{l,k_2} = 0$.
Owing to properties *(P1)* and *(P2)* we obtain

$$E\left[\psi_{k_1,j}\xi_{l,k_2}\right] = E\left[H_j(\{u_i\}_{i=-\infty}^{k_1})\xi_{l,k_2}\right] =$$

$$= EH_j(\{u_i\}_{i=-\infty}^{k_1})E\xi_{l,k_2} = 0.$$

Property (P4): If the measurement noise z_k and the instrumental variables $\psi_{k,j}$ are bounded (i.e. Assumption 5.5 and the condition (C1) are fulfilled), i.e., $|z_k| < z_{\max} < \infty$ and $|\psi_{k,j}| = |H_j(u_k)| < \psi_{\max} < \infty$, then

$$\frac{1}{N}\sum_{k=1}^N \psi_k z_k \to E\psi_k z_k \tag{5.30}$$

with probability 1, as $N \to \infty$; (cf. condition (C3)).
The product $s_{k,j} = \psi_{k,j}z_k$ of stationary and bounded signals $\psi_{k,j}$ and z_k is also stationary, with finite variance (see assumptions of Lemma B.10 on page 225). To prove (5.30), making use of Theorem B.10 in Appendix B.8 we must show, that $r_{s_{k,j}}(\tau) \to 0$, as $|\tau| \to \infty$. Let us notice that the autocovariance function of z_k $(Ez_k = 0)$

$$r_z(\tau) = E\left[(z_k - Ez)(z_{k+\tau} - Ez)\right] = Ez_k z_{k+\tau}, \tag{5.31}$$

as the output of linear filter excited by a white noise has the property that

$$r_z(\tau) \to 0 \tag{5.32}$$

as $|\tau| \to \infty$. Hence, the processes $\psi_{k,j} = H_j(\{u_i\}_{-\infty}^k)$ are ergodic (see *(P3)*), and independent of z_k (see Assumption 5.5). Thus

$$r_{s_{k,j}}(\tau) = E\left[(s_{k,j} - Es_{k,j})(s_{k+\tau,j} - Es_{k,j})\right] \tag{5.33}$$

$$= E\left[\psi_{k,j}\psi_{k+\tau,j}z_k z_{k+\tau}\right] = cr_z(\tau),$$

where $c = (E\psi_{k,j})^2$ is finite constant, $0 \le c < \infty$. Consequently

$$r_{s_{k,j}}(\tau) \to 0 \tag{5.34}$$

as $|\tau| \to \infty$ and

$$\frac{1}{N} \sum_{k=1}^{N} s_{k,j} \to E s_{k,j} \qquad (5.35)$$

with probability 1, as $N \to \infty$.

Property (P5a): For the NARMAX system with the characteristic $\eta()$ as in Assumption 5.7 and the order of autoregression $p = 1$ (see equation (5.1)) it holds that

$$\frac{1}{N} \sum_{k=1}^{N} \psi_k \phi_k^T \to E \psi_k \phi_k^T, \qquad (5.36)$$

with probability 1 as $N \to \infty$, where ψ_k is given by (5.29); compare the condition (C2). For $p = 1$ (for clarity of presentation let also $\lambda_1 = 1$) the system is described by

$$y_k = \eta(y_{k-1}) + \sum_{i=0}^{n} \gamma_i \mu(u_{k-i}) + z_k, \qquad (5.37)$$

and the nonlinearity $\eta()$, according to Assumption 5.7, fulfills the condition

$$|\eta(y)| \le a|y|, \qquad (5.38)$$

where $0 < a < 1$. Introducing the symbol

$$\delta_k = \sum_{i=0}^{n} \gamma_i \mu(u_{k-i}) + z_k, \qquad (5.39)$$

we get

$$y_k = \eta(y_{k-1}) + \delta_k. \qquad (5.40)$$

Since the input $\{u_k\}$ is the i.i.d. sequence, independent of $\{z_k\}$, and the noise $\{z_k\}$ has the property that $r_z(\tau) \to 0$, as $|\tau| \to \infty$ (see (5.32)), we conclude that also $r_\delta(\tau) \to 0$, as $|\tau| \to \infty$. Equation (5.40) can be presented in the following form

$$y_k = \delta_k + \eta\{\delta_{k-1} + \eta[\delta_{k-2} + \eta(\delta_{k-3} + ...)]\}. \qquad (5.41)$$

Let us introduce the coefficients c_k defined, for $k = 1, 2, ..., N$, as follows

$$c_k = \frac{\eta(y_k)}{y_k} \qquad (5.42)$$

with $\frac{0}{0}$ treated as 0. From (5.38) we have that

$$|c_k| \le a < 1, \qquad (5.43)$$

and using c_k, the equation (5.41) can be rewritten as follows

$$y_k = \delta_k + c_{k-1} \left(\delta_{k-1} + c_{k-2} \left(\delta_{k-2} + c_{k-3} \left(\delta_{k-3} + ... \right) \right) \right),$$

i.e.,

$$y_k = \sum_{i=0}^{\infty} c_{k,i} \delta_{k-i},$$

where $c_{k,0} \triangleq 1$, and $c_{k,i} = c_{k-1}c_{k-2}...c_{k-i}$. From (5.43) we conclude that

$$|c_{k,i}| < a^i. \tag{5.44}$$

Since for $0 < a < 1$ the sum $\sum_{i=0}^{\infty} a^i$ is finite, from (5.44) we get $\sum_{i=0}^{\infty} |c_{k,i}| < \infty$, and from (5.39) we simply conclude that for $|\tau| \to \infty$ it holds that $r_y(\tau) \to 0$ and $r_{g_l(y_k)}(\tau) \to 0$, where the processes $g_l(y_k)$ $(l = 1, ..., q)$ are elements of the vector ϕ_k. Thus, for the system with the nonlinearity $\eta()$ as in (5.38) the processes $\{y_k\}$ and $\{g_l(y_k)\}$ $(l = 1, ..., q)$ fulfills assumption of the ergodic law of large numbers (see Lemma B.10 in Appendix B.8), and the property (5.36) holds.

Property (P5b): Under Assumption 5.7, the convergence (5.36) takes place also for the system (5.1) with $p \geq 1$.

For any number sequence $\{x_k\}$ let us define the norm

$$\|\{x_k\}\| = \lim_{K \to \infty} \sup_{k > K} |x_k|. \tag{5.45}$$

and let us present the equation (5.1) in the form

$$y_k = \sum_{j=1}^{p} \lambda_j \eta(y_{k-j}) + \delta_k, \tag{5.46}$$

where δ_k is given by (5.39). The proof of property (P5b) (for $p > 1$) is based of the following theorem (see [80], page. 53).

Theorem 5.3. *Let $\{y_k^{(1)}\}$ and $\{y_k^{(2)}\}$ be two different output sequences of the system (5.1) (see also (5.46)), and $\{\delta_k^{(1)}\}$, $\{\delta_k^{(2)}\}$ be respective aggregated inputs (see (5.39)). If (5.22), (5.23), and (5.24) are fulfilled, then*

$$\frac{1}{1+\alpha} \left\| \{\delta_k^{(1)} - \delta_k^{(2)}\} \right\| \leq \tag{5.47}$$

$$\left\| \{y_k^{(1)} - y_k^{(2)}\} \right\|$$

$$\leq \frac{1}{1-\alpha} \left\| \{\delta_k^{(1)} - \delta_k^{(2)}\} \right\|,$$

where the norm $\| \ \|$ is defined in (5.45).

From (5.47) and under conditions (5.22), (5.23), and (5.24) the steady state of the system (5.1) depends only on the steady state of the input $\{\delta_k\}$. The special case of (5.47) is $\delta_k^{(2)} \equiv 0$, in which $\lim_{K \to \infty} \sup_{k>K} \left| y_k^{(2)} \right| = 0$, and

$$\frac{1}{1+\alpha} \left\| \{\delta_k^{(1)}\} \right\| \leq \left\| \{y_k^{(1)}\} \right\| \leq \frac{1}{1-\alpha} \left\| \{\delta_k^{(1)}\} \right\|.$$

The impulse response of the system tends to zero, as $k \to \infty$ and for the i.i.d. input, the autocorrelation function of the output $\{y_k\}$ is such that

$$r_y(\tau) \to 0, \text{ as } |\tau| \to \infty.$$

Moreover, on the basis of (5.3)–(5.5), since the process $\{y_k\}$ is bounded, it has finite moments of any orders and the ergotic theorems holds (see Lemma B.10 and Lemma B.11 in Appendix B.8). In consequence, the convergence (5.36) holds.

The properties *(P5a)* and *(P5b)* (see (5.36), (5.9) and (5.29)) can be rewritten for particular elements of ψ_k and ϕ_k in the following way

$$\frac{1}{N} \sum_{k=1}^{N} \psi_{k_1,j} g_l(y_{k_2}) \to E\psi_{k_1,j} g_l(y_{k_2})$$

with probability 1, as $N \to \infty$.

Under the property that $E[\psi_{k_1,j} \xi_{l,k_2}] = 0$ (see *(P3)*), for instrumental variables generated according to (5.29) we obtain

$$E[\psi_{k_1,j} g_l(y_{k_2})] = E[\psi_{k_1,j} G_{l,k_2}].$$

Denoting (cf. (5.9))

$$\Phi_N^\# = (\phi_1^\#, \phi_2^\#, ..., \phi_N^\#)^T, \tag{5.48}$$
$$\phi_k^\# \triangleq (f_1(u_k), ..., f_m(u_k), ..., f_1(u_{k-n}), ..., f_m(u_{k-n}),$$
$$G_{1,k-1}, ..., G_{q,k-1}, ..., G_{1,k-p}, ..., G_{q,k-p})^T,$$

where $G_{l,k} \triangleq E\{g_l(y_k) \mid \{u_i\}_{i=-\infty}^{k}\}$ (see (5.25)), and making use of ergodicity of the processes $\{\psi_{k,j}\}$ ($j = 1, ..., m(n+1)+pq$), $\{f_t(u_k)\}$ ($t = 1, ..., m$) and $\{G_{l,k}\}$ ($l = 1, ..., q$) (see (5.29) and Assumption 5.1) we get

$$\frac{1}{N} \Psi_N^T \Phi_N^\# = \frac{1}{N} \sum_{k=1}^{N} \psi_k \phi_k^{\#T} \to E\psi_k \phi_k^{\#T} \text{ with p. 1,}$$

and using (5.36) we get

$$\frac{1}{N} \Psi_N^T \Phi_N = \frac{1}{N} \sum_{k=1}^{N} \psi_k \phi_k^T \to E\psi_k \phi_k^T \text{ with p. 1,}$$

for the instruments as in (5.29). Directly from definitions (5.25) and (5.48) we conclude that $E\left[\psi_{k_1,j} g_l(y_{k_2})\right] = E\left[\psi_{k_1,j} G_{l,k_2}\right]$ and

$$E\psi_k \phi_k^{\#T} = E\psi_k \phi_k^T.$$

Thus, for any choice of instrumental variables matrix Ψ_N, which fulfills the property (P3) (see (5.29)), the following equivalence takes place asymptotically with probability 1, as $N \to \infty$

$$\frac{1}{N}\Psi_N^T \Phi_N^{\#} = \frac{1}{N}\Psi_N^T \Phi_N. \tag{5.49}$$

The estimation error (i.e., the difference between the estimate and the true value of parameters) has the form

$$\Delta_N^{(IV)} = \widehat{\theta}_N^{(IV)} - \theta = \left(\frac{1}{N}\Psi_N^T \Phi_N\right)^{-1} \left(\frac{1}{N}\Psi_N^T Z_N\right).$$

Introducing

$$\Gamma_N \triangleq \left(\frac{1}{N}\Psi_N^T \Phi_N\right)^{-1} \frac{1}{\sqrt{N}}\Psi_N^T,$$

$$Z_N^* \triangleq \frac{\frac{1}{\sqrt{N}} Z_N}{z_{\max}},$$

where $z_{\max}$ upper bound of the absolute value of the noise (see Assumption 5.5) we obtain

$$\Delta_N^{(IV)} = z_{\max} \Gamma_N Z_N^*. \tag{5.50}$$

with the Euclidean norm of Z_N^*

$$\|Z_N^*\| = \sqrt{\sum_{k=1}^N \left(\frac{\frac{1}{\sqrt{N}} z_k}{z_{\max}}\right)^2} = \sqrt{\frac{1}{N}\sum_{k=1}^N \left(\frac{z_k}{z_{\max}}\right)^2} \le 1.$$

Let the quality of the instrumental variables be evaluated on the basis of the following criterion (see e.g. [154])

$$Q(\Psi_N) = \max_{\|Z_N^*\| \le 1} \left\|\Delta_N^{(IV)}(\Psi_N)\right\|^2, \tag{5.51}$$

where $\|\ \|$ denotes the Euclidean norm, and $\Delta_N^{(IV)}(\Psi_N)$ is the estimation error obtained for the instrumental variables Ψ_N.

Theorem 5.4. *If the Assumptions 5.3–5.7 and the condition (5.29) hold, then the criterion $Q(\Psi_N)$ given by (5.51) attains minimum for the choice*

$$\Psi_N^\# = \Phi_N^\# \qquad (5.52)$$

i.e., for each Ψ_N and N large it holds that

$$Q(\Psi_N^\#) \leqslant Q(\Psi_N) \text{ with } p. \ 1.$$

Proof. For the proof see Appendix A.21.   ∎

Obviously, instrumental variables given by (5.52) fulfill postulates (C1)÷(C3).

5.6   Nonparametric Generation of Instrumental Variables

The optimal matrix of instruments $\Psi_N^\#$ cannot be computed analytically, because of the lack of prior knowledge of the system (the probability density functions of excitations and the values of parameters are unknown). Estimation of $\Psi_N^\#$ is also difficult, because the elements $G_{l,k}$ depends on infinite number of measurements of the input process. Therefore, we propose the following heuristic method

$$\Psi_N^{(r)\#} = (\psi_1^{(r)\#}, \psi_2^{(r)\#}, ..., \psi_N^{(r)\#})^T,$$
$$\psi_k^{(r)\#} \triangleq (f_1(u_k), ..., f_m(u_k), ..., f_1(u_{k-n}), ..., f_m(u_{k-n}),$$
$$G_{1,k-1}^{(r)}, ..., G_{q,k-1}^{(r)}, ..., G_{1,k-p}^{(r)}, ..., G_{q,k-p}^{(r)})^T$$

where

$$G_l^{(r)} = G_l^{(r)}(u^{(0)}, ..., u^{(r)}) \qquad (5.53)$$
$$\triangleq E\{g_l(y_j) \mid u_j = u^{(0)}, ..., u_{j-r} = u^{(r)}\},$$
$$G_{l,k}^{(r)} = G_l^{(r)}(u_k, ..., u_{k-r}).$$

It is based on the intuition that the approximate value $\Psi_N^{(r)\#}$ becomes better, i.e.,

$$\Psi_N^{(r)\#} \cong \Psi_N^\#,$$

when r grows (this question is treated as open). For $r = 0$ we have

$$\Psi_N = \Psi_N^{(0)\#},$$
$$\psi_k^{(0)\#} \triangleq (f_1(u_k), .., f_m(u_k), .., f_1(u_{k-n}), .., f_m(u_{k-n}),$$
$$R_1(u_{k-1}), .., R_q(u_{k-1}), .., R_1(u_{k-p}), .., R_q(u_{k-p}))^T,$$

where

$$R_l(u) = G_l^{(0)}(u) = E\{g_l(y_k)\} \mid u_k = u\}. \qquad (5.54)$$

All elements of $\psi_k^{(0)\#}$ (white noises) fulfill (P3). After introducing

$$x_{l,k} = g_l(y_k),$$

the regression functions in (5.54) can be written as

$$R_l(u) = E\{x_{l,k} \mid u_k = u\}.$$

Both u_k and y_k can be measured, and $x_{l,k} = g_l(y_k)$ can be computed, because the functions $g_l()$ are known a priori. The most natural method for generation of $\Psi_N^{(r)\#}$ is thus the kernel method. Traditional estimate of the regression function $R_l(u)$ computed on the basis of M pairs $\{(u_i, x_{l,i})\}_{i=1}^M$ has the form (see e.g. [50])

$$\widehat{R}_{l,M}(u) = \frac{\frac{1}{M}\sum_{i=1}^M \left(x_{l,i}K\left(\frac{u-u_i}{h(M)}\right)\right)}{\frac{1}{M}\sum_{i=1}^M K\left(\frac{u-u_i}{h(M)}\right)}, \tag{5.55}$$

where $K()$ is a kernel function, and $h()$ – the bandwidth parameter. Further considerations will be based on the following two theorems, (see [50]).

Theorem 5.5. *If $h(M) \to 0$ and $Mh(M) \to \infty$ as $M \to \infty$, and $K(v)$ is one of $\exp(-|v|)$, $\exp(-v^2)$, or $\frac{1}{1+|v|^{1+\delta}}$, then*

$$\frac{\frac{1}{M}\sum_{i=1}^M \left(y_i K\left(\frac{u-u_i}{h(M)}\right)\right)}{\frac{1}{M}\sum_{i=1}^M K\left(\frac{u-u_i}{h(M)}\right)} \to E\{y_i \mid u_i = u\} \tag{5.56}$$

in probability as $M \to \infty$, provided that $\{(u_i, y_i)\}_{i=1}^M$ is an i.i.d. sequence.

Theorem 5.6. *If both the regression $E\{y_i \mid u_i = u\}$, and the input probability density function $\vartheta(u)$ have finite second order derivatives, then for $h(M) = O(M^{-\frac{1}{5}})$ the asymptotic rate of convergence in (5.56) is $O(M^{-\frac{2}{5}})$ in probability.*

To apply above theorems, let us additionally take the following assumption.

Assumption 5.8. *The functions $g_1(y),...,g_q(y)$, $f_1(u),...,f_m(u)$ and the input probability density $\vartheta(u)$ have finite second order derivatives for each $u \in (-u_{\max}, u_{\max})$ and each $y \in (-y_{\max}, y_{\max})$.*

In our problem, the process $\{x_{l,i}\}$ appearing in the numerator of (5.55) is correlated. Let us decompose the sums in numerator and denominator in (5.55) for $r = \left\lfloor M^{\frac{1}{\chi(M)}} \right\rfloor$ partial sums, where $\chi(M)$ is such that $\chi(M) \to \infty$ and $r \to \infty$, as $M \to \infty$ (e.g. $\chi(M) = \sqrt{\log M}$), i.e.

$$L(\{(u_i, x_{l,i})\}_{i=1}^M) \tag{5.57}$$

$$\triangleq \frac{1}{M} \sum_{i=1}^M x_{l,i} K\left(\frac{u - u_i}{h(M)}\right) = \frac{1}{r} \sum_{t=1}^r s_t,$$

$$W(\{u_i\}_{i=1}^M) \triangleq \frac{1}{M} \sum_{i=1}^M K\left(\frac{u - u_i}{h(M)}\right) = \frac{1}{r} \sum_{t=1}^r w_t,$$

with

$$s_t = \frac{1}{\frac{M}{r}} \sum_{\{i:0<ir+t\le M\}} x_{l,ir+t} K\left(\frac{u - u_{ir+t}}{h(M)}\right), \tag{5.58}$$

$$w_t = \frac{1}{\frac{M}{r}} \sum_{\{i:0<ir+t\le M\}} K\left(\frac{u - u_{ir+t}}{h(M)}\right).$$

The components of the sum (5.57) have the time distance r and become uncorrelated as $r \to \infty$. This fact is a simple consequence of the property that $r_x(\tau) \to 0$, as $|\tau| \to \infty$. Moreover, the components in (5.58) are i.i.d. Each of the sub-sums $\{s_t\}$ has the same probability density, but uses different subset of measurements. All of them includes $\overline{M} = \frac{M}{r}$ data. For simplicity let us write

$$s_t = \frac{1}{\overline{M}} \sum_{\{i:0<ir+t\le M\}} x_{l,ir+t} K\left(\frac{u - u_{ir+t}}{H(\overline{M})}\right), \tag{5.59}$$

$$w_t = \frac{1}{\overline{M}} \sum_{\{i:0<ir+t\le M\}} K\left(\frac{u - u_{ir+t}}{H(\overline{M})}\right),$$

where $H(\overline{M}) \triangleq h(M)$. Let $h(M) = cM^\alpha$, where $-1 < \alpha < 0$, then

$$H(\overline{M}) = cM^\alpha = c\left(M^{\frac{1 - \frac{1}{\chi(M)}}{1 - \frac{1}{\chi(M)}}}\right)^\alpha \tag{5.60}$$

$$= c\left(\overline{M}^\alpha\right)^{\frac{1}{1 - \frac{1}{\chi(M)}}} = O(\overline{M}^\alpha)$$

and for $\overline{M} \to \infty$, it holds that

$$H(\overline{M}) \to 0 \text{ and } \overline{M}H(\overline{M}) \to \infty. \tag{5.61}$$

From (5.59), (5.60), (5.61) and Theorem 5.5, for $r \to \infty$ we get

$$P \lim_{\overline{M} \to \infty}\left(\frac{s_t}{w_t}\right) = \frac{P \lim_{\overline{M} \to \infty}(s_t)}{P \lim_{\overline{M} \to \infty}(w_t)} = \frac{a(u)}{b(u)} = R_l(u),$$

for each $t = 1, 2, ..., r$, and since

$$\widehat{R}_{l,M}(u) = \frac{L(\{(u_i, x_{l,i})\}_{i=1}^M)}{W(\{u_i\}_{i=1}^M)} = \frac{\frac{1}{r}\sum_{t=1}^r s_t}{\frac{1}{r}\sum_{t=1}^r w_t},$$

we obtain that

$$P \lim_{M \to \infty} \left(\widehat{R}_{l,M}(u)\right) = R_l(u). \tag{5.62}$$

Under Assumption 5.8, from the property (5.60) and Theorem 5.6 we conclude that for $h(M) = cM^{-\frac{1}{5}}$ the rate of convergence of (5.55) is $O(M^{-\frac{2}{5}})$ in probability.

5.7 The 3-Stage Identification

Taking into account the conclusions from Section 5.6, in particular the form of optimal instruments Ψ_N^*, we propose the following combined parametric-nonparametric identification procedure (see [91], and [92]).

Stage 1 (nonparametric): Using $M + \max(n, p)$ measurements

$$\{(u_i, y_i)\}_{i=1-\max(n,p)}^M$$

generate empirical matrix of instruments $\widehat{\Psi}_{N,M}^* = (\widehat{\psi}_{1,M}^*, \widehat{\psi}_{2,M}^*, ...\widehat{\psi}_{N,M}^*)^T$, where

$$\widehat{\psi}_{k,M}^* = (f_1(u_k), ..., f_m(u_k), ..., f_1(u_{k-n}), ... \tag{5.63}$$
$$..., f_m(u_{k-n}), \widehat{R}_{1,M}(u_{k-1}), ...,$$
$$\widehat{R}_{q,M}(u_{k-1}), ..., \widehat{R}_{1,M}(u_{k-p}), ..., \widehat{R}_{q,M}(u_{k-p}))^T,$$

and $\widehat{R}_{l,M}(u) = \sum_{i=1}^M \left(g_l(y_i)K(\frac{u-u_i}{h(M)})\right) / \sum_{i=1}^M K(\frac{u-u_i}{h(M)})$.

Stage 2 (parametric): Estimate the aggregated parameter vector (5.8)

$$\theta = (\gamma_0 c_1, ..., \gamma_0 c_m, .., \gamma_n c_1, ..., \gamma_n c_m, \lambda_1 d_1, ..., \lambda_1 d_q, ..., \lambda_p d_1, ..., \lambda_p d_q)^T$$

by the instrumental variables method

$$\widehat{\theta}_{N,M}^{*(IV)} = \left(\widehat{\Psi}_{N,M}^{*T} \Phi_N\right)^{-1} \widehat{\Psi}_{N,M}^{*T} Y_N, \tag{5.64}$$

where

$$Y_N = (y_1, y_2, ..., y_N)^T, \qquad \Phi_N = (\phi_1, \phi_2, ..., \phi_N)^T,$$
$$\phi_k = (f_1(u_k), ..., f_m(u_k), ..., f_1(u_{k-n}), ..., f_m(u_{k-n}),$$
$$g_1(y_{k-1}), ..., g_q(y_{k-1}), ..., g_1(y_{k-p}), ..., g_q(y_{k-p}))^T,$$

(see (5.9)), and next, using $\widehat{\theta}_{N,M}^{*(IV)}$ construct the estimates $\widehat{\Theta}_{\lambda d}^{(IV)}$ and $\widehat{\Theta}_{\gamma c}^{(IV)}$ of the matrices $\Theta_{\lambda d} = \Lambda d^T$ and $\Theta_{\gamma c} = \Gamma c^T$.

Stage 3 (decomposition): Compute the SVD (singular value decomposition) of the matrices $\widehat{\Theta}_{\lambda d}^{(IV)}$ and $\widehat{\Theta}_{\gamma c}^{(IV)}$, i.e., $\widehat{\Theta}_{\gamma c}^{(IV)} = \sum_{i=1}^{\min(n,m)} \sigma_i \widehat{\mu}_i \widehat{\nu}_i^T$, $\widehat{\Theta}_{\lambda d}^{(IV)} = \sum_{i=1}^{\min(p,q)} \delta_i \widehat{\xi}_i \widehat{\zeta}_i^T$ to obtain the estimates of the parameters (elements of the impulse responses of the linear dynamic blocks and the parameters of static nonlinear characteristics)

$$\widehat{\Lambda}_N = sgn(\widehat{\xi}_1[\kappa_{\xi_1}])\widehat{\xi}_1, \qquad \widehat{\Gamma}_N = sgn(\widehat{\mu}_1[\kappa_{\mu_1}])\widehat{\mu}_1, \qquad (5.65)$$
$$\widehat{c}_N = sgn(\widehat{\mu}_1[\kappa_{\mu_1}])\sigma_1 \widehat{\nu}_1, \qquad \widehat{d}_N = sgn(\widehat{\xi}_1[\kappa_{\xi_1}])\delta_1 \widehat{\zeta}_1,$$

where $x[k]$ denotes k-th element of the vector x, and $\kappa_{\mathbf{x}} = \min\{k : x[k] \neq 0\}$. Under condition (5.62) the following theorem holds.

Theorem 5.7. *For the NARMAX system with the characteristic $\eta(y)$ as in Assumption 5.7 it holds that*

$$\widehat{\theta}_{N,M}^{*(IV)} \to \theta, \text{ in probability}$$

as $M \to \infty$ and $N \to \infty$, provided that $h(M)$ fulfills assumptions of Theorem 5.5.

Proof. For the proof see the Appendix A.22.   ∎

5.8 Example

5.8.1 Simulation

To illustrate behaviour of the method, we simulated a special case of the general NARMAX model (5.1), commonly known in the literature as a Lur'e system (see Fig. 5.2), and often met in applications (see [72], [73], [85], [135]). In this case, the static block $\mu()$ is linear, i.e., $\mu(u) = u$, and both linear dynamic blocks $\{\gamma_i\}$ and $\{\lambda_j\}$ have the same impulse responses. Thus, in the computer experiment we set

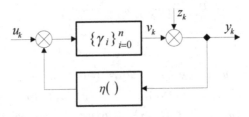

Fig. 5.2 The Lur'e system

$$n = 3,$$
$$\gamma_0 = 0, \qquad \gamma_1 = 1, \qquad \gamma_2 = 1,$$
$$p = 2,$$
$$\lambda_j = \gamma_j, \qquad j = 1, 2,$$

and we applied the nonlinear feedback of the form

$$\eta(y) = \frac{1}{4} |y|.$$

Since, for the considered case

$$r = \frac{1}{4}, \qquad \sum_{j=1}^{p} \lambda_j = 2,$$

$$\alpha = r \sum_{j=1}^{p} \lambda_j = \frac{1}{2} < 1,$$

the simulated system is stable (see (5.24)), and can be described by the following nonlinear difference equation

$$y_k = u_{k-1} + u_{k-2} + \frac{1}{4} |y_{k-1}| + \frac{1}{4} |y_{k-2}| + z_k.$$

The system was excited by the uniformly distributed random sequence $u_k \sim U[-1, 1]$, and disturbed by the colored noise $z_k = \frac{1}{2} z_{k-1} + \varepsilon_k$, where $\varepsilon_k \sim U[-1, 1]$.

5.8.2 Identification

We assumed linear model of $\mu()$

$$\mu(u) = c_1 u + c_2,$$
$$\text{i.e., } f_1(u) = u, f_2(u) = 1, m = 2$$

and two-segment piecewise linear model of $\eta()$

$$\eta(y) = d_1 y \cdot 1(y) + d_2 y \cdot 1(-y),$$
$$\text{i.e., } g_1(y) = y \cdot 1(y), g_2(y) = y \cdot 1(-y), q = 2,$$
$$\text{where } 1(x) = \begin{cases} 1, & \text{as } x \geq 0, \\ 0, & \text{elsewhere.} \end{cases}$$

The system with the true vectors of parameters

$$\Lambda_{\text{true}} = (1,1)^T, \qquad \Gamma_{\text{true}} = (0,1,1)^T,$$

$$c_{\text{true}} = (1,0)^T, \qquad d_{\text{true}} = \left(\frac{1}{4}, -\frac{1}{4}\right)^T,$$

was normalized to the following equivalent version (see condition (b) on page 115)

$$\Lambda = \left(\frac{\sqrt{2}}{2}, \frac{\sqrt{2}}{2}\right)^T, \qquad \Gamma = \left(0, \frac{\sqrt{2}}{2}, \frac{\sqrt{2}}{2}\right)^T,$$

$$c = (\sqrt{2}, 0)^T, \qquad d = \left(\frac{\sqrt{2}}{4}, -\frac{\sqrt{2}}{4}\right)^T.$$

The aggregated vector of mixed products of parameters θ and identified matrices $\Theta_{\Lambda d}$ and $\Theta_{\Gamma c}$ are as follows

$$\theta = \left(0, 0, 1, 0, 1, 0, \frac{1}{4}, -\frac{1}{4}, \frac{1}{4}, -\frac{1}{4}\right)^T,$$

$$\Theta_{\Lambda d} = \begin{bmatrix} \frac{1}{4} & -\frac{1}{4} \\ \frac{1}{4} & -\frac{1}{4} \end{bmatrix}, \qquad \Theta_{\Gamma c} = \begin{bmatrix} 0 & 0 \\ 1 & 0 \\ 1 & 0 \end{bmatrix}.$$

We compared the estimates (5.11) and (5.19) with

$$\phi_k = (u_k, 1, u_{k-1}, 1, u_{k-2}, 1, y_{k-1}1(y_{k-1}),$$
$$y_{k-1}1(-y_{k-1}), y_{k-2}1(y_{k-2}), y_{k-2}1(-y_{k-2}))^T,$$
$$\widehat{\psi}^*_{k,M} = \left(u_k, 1, u_{k-1}, 1, u_{k-2}, 1, \widehat{R}_{1,M}(u_{k-2}),\right.$$
$$\left. \widehat{R}_{2,M}(u_{k-2}), \widehat{R}_{1,M}(u_{k-3}), \widehat{R}_{2,M}(u_{k-3})\right)^T,$$

where

$$\widehat{R}_{l,M}(u) = \frac{\frac{1}{M}\sum_{i=1}^{M}\left(g_l(y_{i+1})K\left(\frac{u-u_i}{h(M)}\right)\right)}{\frac{1}{M}\sum_{i=1}^{M}K\left(\frac{u-u_i}{h(M)}\right)}.$$

The Mean Normalized Errors of both subsystems

$$MNE_\Gamma = \frac{\left\|\widehat{\Gamma}_N - \Gamma\right\|_2}{\|\Gamma\|_2},$$

$$MNE_d = \frac{\left\|\widehat{d}_N - d\right\|_2}{\|d\|_2},$$

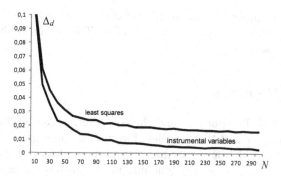

Fig. 5.3 Estimation error of the nonlinear static block

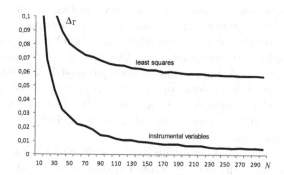

Fig. 5.4 Estimation error of the linear dynamic block

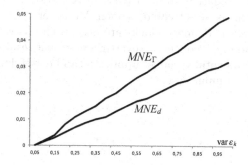

Fig. 5.5 Estimation error vs. variance of the noise, for $N = 100$ (instrumental variables method)

were computed and averaged over 10 re-runs, and for various numbers of measurements. Figures 5.3 and 5.4 show that on the contrary to the least squares method, the algorithm is free of asymptotic bias (i.e. as $N \to \infty$) and converges to true system parameters. The experiment was also repeated for various variances of the noise ε_k. The results for $N = 100$, shown in

Fig. 5.5, confirm linear increase of the estimation errors, which is typical in 'linear in the parameters' system identification. The results confirm usability of the proposed scheme.

5.9 Final Remarks

The advantages of the approach and the contribution can be summarized as follows. The structure of the considered system is more general than Hammerstein systems and Lur'e systems. Moreover, nonlinear characteristics of static blocks need not to be of polynomial form, which is commonly assumed in the literature. Also excitations can have arbitrary correlation properties. The method, as a whole, is computationally simple, and standard numerical LS/IV procedures (e.g. LU and Cholesky decomposition) can be applied in the main stage of the routine. The consistency of the proposed estimate is proved, even for correlated noise, and with correlation between the noise and the input, caused by structural feedback. Full versions of the proofs of theorems are included. Good cooperation between parametric and nonparametric methods is shown. The problem of suboptimal generation of instrumental variables is solved by application of nonparametric (kernel) methods. Also the scope of applicability of instrumental variables technique is extended for nonlinear systems with feedback.

Obviously, the proposed algorithm has some drawbacks. The most significant is the fact that the class is limited to the 'linear in the parameters' additive NARMAX models, and neither input cross-terms, nor lagged noise terms are admitted in the difference equation describing the system. The consistency of the estimate with intuitive approximation $\Psi_N^{(r)}$ of Ψ_N was not proved formally. This issue is treated as open. Moreover, for technical reasons (SVD method), only FIR linear blocks are acceptable. It was also assumed that the input is an i.i.d. sequence. Nevertheless, recent results (see e.g. [98]) show that the instrumental variables approach can be useful for reducing the bias in the correlated input case.

Chapter 6
Large-Scale Interconnected Systems

This chapter addresses the problem of parameter estimation of elements in complex, interconnected systems. Similarity between causes of biases in the least squares estimates for a simple *SISO linear dynamic* object, and for a *MIMO linear static* system with composite structure, was noticed. For linear complex static system, the instrumental variable estimate is proposed and compared with the least squares approach. The strong consistency of the presented parameter estimate is proved. Also the optimal values of instrumental variables are established, and the method of their suboptimal generation is presented. The conclusions are verified in numerical experiments.

6.1 Introduction

We consider the problem of parameter estimation in complex, interconnected systems with the presence of random noises. In a lot of commonly met hierarchical control problems (see [32]), the accurate mathematical models of the particular system components are needed. Under the term 'complex' we understand the fact that the system is built of a number of interconnected components (subsystems), e.g., in the typical production system each element is excited by the outputs of other blocks (see [32]). In consequence of mutual interconnections, the components are dependent and their separation may be impossible or too expensive. In general, excitations of particular element cannot be freely generated in the experiment. It leads to the problem of structural identifiability (i.e. identifiability of separate elements does not imply identifiability of the whole interconnected system (see [61] and [62])) and usually badly conditioned numerical tasks. Moreover, some interaction signals are hidden, and cannot be directly measured. For these reasons, the algorithms dedicated for single element cannot be directly applied in complex system analysis.

Identifiability of the element, which operates in complex system, depends additionally on the system structure and even on the values of parameters of

G. Mzyk, *Combined Parametric-Nonparametric Identification of Block-Oriented Systems*, Lecture Notes in Control and Information Sciences 454, DOI: 10.1007/978-3-319-03596-3_6, © Springer International Publishing Switzerland 2014

other elements. Particularly, the components preceding identified object must guarantee persistency of the input excitation [61]. We apply and compare two methods – least squares (*l.s.*) and instrumental variables (*i.v.*).

It is commonly known from the linear system theory, that the least squares approach applied for the simple SISO linear dynamic object leads to biased estimate. The reason of the bias results from the property of autoregression, i.e. the correlation between the noise and the values of previous outputs of the identified object (see the Appendix A.23). Analogously, for the complex, interconnected systems with random noises, the least squares estimate has the non-zero systematic error even if the number of measurement data tends to infinity. The reason of the bias is that the output noises are transferred to the inputs through the structural feedback.

In the chapter, the formal similarity of these problems is shown and the instrumental variables technique, used so far for the linear dynamics identification, was successfully generalized for the systems with complex structure. It is shown that the proposed *i.v.* estimate is strongly consistent independently of the system structure and the color of the noise. Moreover, the computational complexity of the method is comparable with the *l.s.*algorithm. In Section 6.2 the identification problem and the purpose is formulated in detail. Next, in Section 6.3, the properties of the least squares based algorithm proposed in [62] are reminded. In particular, the reason of its bias is shown in detail, and in Section 6.4 the new *i.v.* estimate is introduced and analyzed. The model is generalized for nonlinear dynamic block in Section 6.5, and finally, in Section 6.6, the performance of the method is demonstrated by the simulation example.

6.2 Statement of the Problem

Consider the system shown in Fig. 6.1. It consists of n linear elements described as follows

$$y_i = a_i x_i + b_i u_i + \xi_i \qquad (i = 1, 2, ..., n),$$

where

$$u = (u_1, u_2, ..., u_n)^T$$
$$x = (x_1, x_2, ..., x_n)^T$$
$$y = (y_1, y_2, ..., y_n)^T$$

are the external inputs, interaction inputs, and system outputs, respectively. The processes

$$\delta = (\delta_1, \delta_2, ..., \delta_n)^T$$
$$\xi = (\xi_1, \xi_2, ..., \xi_n)^T$$

are random disturbances. The block H determines known structure of inter-connections in the following way

$$x_i = H_i y + \delta_i, \qquad (6.1)$$

where H_i is the ith row of the binary matrix H (i.e. $H_{i,j} = 0$ – 'no connection', $H_{i,j} = 1$ – 'connection').

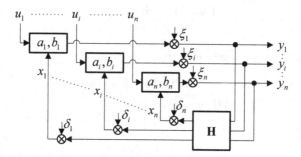

Fig. 6.1 The complex n-element linear static system

The aim is to estimate parameters $\{(a_i, b_i)\}_{i=1}^n$ of the particular elements using the set of data $\{(u^{(k)}, y^{(k)})\}_{k=1}^N$ collected in the experiment. We emphasize that the internal excitations $x^{(k)}$ cannot be measured.

We assume that:

Assumption 6.1. *The structure of the system (i.e., the matrix H) is known.*

Assumption 6.2. *The system is well defined, i.e., for any (u, δ, ξ) it exists unique y (see [62]).*

Assumption 6.3. *The noises δ, and ξ are zero-mean, mutually independent, and independent of u.*

Assumption 6.4. *In the noise-free case ($\delta = 0$, and $\xi = 0$) the system would be identifiable (see [62]).*

Assumption 6.5. *The excitations are rich enough, i.e., the matrix*

$$E_N = (e_1, e_2, ..., e_N),$$

where

$$e = (u^T, \theta^T)^T \text{ and } \theta = A\delta + \xi,$$

is of full rank with probability 1.

Introducing the matrices

$$A = \text{diag}(a_1, a_2, ..., a_n) \tag{6.2}$$
$$B = \text{diag}(b_1, b_2, ..., b_n) \tag{6.3}$$
$$H = \left(H_1^T, H_2^T, ..., H_n^T\right)^T \tag{6.4}$$

the whole system can be described in the following compact form

$$\begin{cases} y = Ax + Bu + \xi \\ x = Hy + \delta \end{cases}. \tag{6.5}$$

Inserting x to the first equation in (6.5) we obtain

$$y = A(Hy + \delta) + Bu + \xi,$$
$$(I - AH)y = Bu + A\delta + \xi,$$

which leads to

$$y = Ku + G\theta, \tag{6.6}$$

where

$$G = (I - AH)^{-1}, \tag{6.7}$$
$$K = (I - AH)^{-1}B = GB.$$

The equation (6.6) resembles description of the object with the input u, the output y, the transfer matrix K, and the noise $G\theta$. Invertibility of $(I - AH)$ in (6.7) is equivalent to Assumption 6.2.

6.3   Least Squares Approach

Introducing the vectors of input-output data of ith element

$$Y_{iN} = [y_i^{(1)}, y_i^{(2)}, ..., y_i^{(N)}], \tag{6.8}$$
$$W_{iN} = [w_i^{(1)}, w_i^{(2)}, ..., w_i^{(N)}],$$
$$\text{where } w_i = (x_i, u_i)^T,$$

we obtain the measurement equation

$$Y_{iN} = (a_i, b_i)W_{iN} + \xi_{iN}. \tag{6.9}$$

Since the input x_i included in w_i is unknown (cannot be measured), the least squares estimate cannot be derived directly from (6.9). Owing to (6.1) the natural substitution is

$$\tilde{w}_i = (\tilde{x}_i, u_i)^T, \text{ where } \tilde{x}_i = H_i y = x_i - \delta_i.$$

It leads to the following least squares estimate

$$(\widehat{a}_i^{l.s.}, \widehat{b}_i^{l.s.}) = Y_{iN} \widetilde{W}_{iN}^T \left(\widetilde{W}_{iN} \widetilde{W}_{iN}^T \right)^{-1}, \tag{6.10}$$

where

$$\widetilde{W}_{iN} = [\widetilde{w}_i^{(1)}, \widetilde{w}_i^{(2)}, ..., \widetilde{w}_i^{(N)}].$$

Remark 6.1. *The estimate (6.10) originates from the modified version of measurement equation (6.9), in which W_{iN} was substituted with $\widetilde{W}_{iN}$*

$$Y_{iN} = (a_i, b_i)\widetilde{W}_{iN} + \Theta_{iN}. \tag{6.11}$$

Consequently, in (6.11) the disturbance

$$\Theta_{iN} = [\theta_i^{(1)}, \theta_i^{(1)}, ..., \theta_i^{(N)}]$$

appears instead of ξ_{iN}. The situation is similar to the problem of identification of the simple linear dynamics with autoregression (see example in the Appendix A.23). It was shown in [62], that because of correlation between the elements of Θ_{iN} and $\widetilde{W}_{iN}$, the estimation error

$$(\widehat{a}_i^{l.s.}, \widehat{b}_i^{l.s.}) - (a_i, b_i) = \Theta_{iN} \widetilde{W}_{iN} \left(\widetilde{W}_{iN} \widetilde{W}_{iN}^T \right)^{-1}$$

does not tend to zero, as $N \to \infty$.

6.4 Instrumental Variables Approach

To solve the problem mentioned in Remark 6.1 we propose the analogous strategy as for the SISO dynamic system identification (see e.g. [46] and [92]), i.e., generalization of (6.10) to the following form

$$(\widehat{a}_i^{i.v.}, \widehat{b}_i^{i.v.}) = Y_{iN} \Psi_{iN}^T \left(\widetilde{W}_{iN}^T \Psi_{iN}^T \right)^{-1}, \tag{6.12}$$

where

$$\Psi_{iN} = [\psi_i^{(1)}, \psi_i^{(2)}, ..., \psi_i^{(N)}]$$

is the additional matrix of instrumental variables, of the same dimensions as $\widetilde{W}_{iN}$, i.e.,

$$\psi_i^{(k)} = \left(\psi_{i,1}^{(k)}, \psi_{i,2}^{(k)} \right)^T, \qquad \dim \psi_{i,1} = \dim x_i, \qquad \dim \psi_{i,2} = \dim u_i.$$

We impose the following two conditions on Ψ_{iN} (see e.g. [96]):

(C1) The instrumental variables $\psi_{i,1}$, and $\psi_{i,2}$ are correlated with the input u_i, such that

$$\frac{1}{N}\widetilde{W}_{iN}^T \Psi_{iN}^T = \frac{1}{N}\sum_{k=1}^{N}\psi_i^{(k)}\widetilde{w}_i^{(k)^T} \to E\psi_i\widetilde{w}_i^T$$

with probability 1 as $N \to \infty$, and the limit matrix $E\psi_i\widetilde{w}_i^T$ is of full rank.

(C2) Simultaneously, $\psi_{i,1}$, and $\psi_{i,2}$ are *not* correlated with the aggregated output noise θ_i, i.e.,

$$\Psi_{iN} = L_i E_N,$$

where

$$L_i = \begin{bmatrix} \Gamma_i & 0 \\ I_i & 0 \end{bmatrix},$$

and

$$I_i = [0, ..., 0, 1, 0, ..., 0].$$

The following theorem holds.

Theorem 6.1. *If the instrumental variables matrix Ψ_{iN} fulfils* (C1) *and* (C2) *then*

$$(\widehat{a}_i^{i.v.}, \widehat{b}_i^{i.v.}) \to (a_i, b_i) \tag{6.13}$$

with probability 1, *as $N \to \infty$*.

Proof. The estimation error has the form

$$\Delta = (\widehat{a}_i^{i.v.}, \widehat{b}_i^{i.v.}) - (a_i, b_i) = \tag{6.14}$$

$$= \Theta_{iN}\Psi_{iN}^T \left(\widetilde{W}_{iN}\Psi_{iN}^T\right)^{-1} = \left(\frac{1}{N}\Theta_{iN}\Psi_{iN}^T\right)\left(\frac{1}{N}\widetilde{W}_{iN}\Psi_{iN}^T\right)^{-1}$$

where

$$\widetilde{W}_{iN} = F_i E_N$$

and

$$F_i = \begin{bmatrix} H_i K & H_i G \\ I_i & 0 \end{bmatrix}.$$

Since, according to *(C1)* and *(C2)* it holds that

$$\frac{1}{N}\widetilde{W}_{iN}\Psi_{iN}^T \to \begin{bmatrix} \mathrm{cov}(\widetilde{x}_i, \psi_{i,1}) & \mathrm{cov}(u_i, \psi_{i,1}) \\ \mathrm{cov}(\widetilde{x}_i, \psi_{i,2}) & \mathrm{cov}(u_i, \psi_{i,2}) \end{bmatrix},$$

$$\frac{1}{N}\Theta_{iN}\Psi_{iN}^T \to 0,$$

from the Slutzky theorem we conclude (6.13). ∎

In real applications the procedure of Ψ_{iN}-generation is of fundamental meaning. Let us introduce the quality index of instrumental variables

$$Q(\Psi_{iN}) = \|\Delta(\Psi_{iN})\| = \lambda_{\max}\left(\Delta(\Psi_{iN})\Delta^T(\Psi_{iN})\right). \tag{6.15}$$

The following theorem holds.

Theorem 6.2. *The optimal instruments with respect to the value of* $Q(\Psi_{iN})$ *has the form*

$$\psi_i^* = \overline{w}_i = (\overline{x}_i, u_i)^T, \text{ where } \overline{x}_i = E(x_i|u) = H_i K u, \qquad (6.16)$$

i.e.,

$$\Gamma_i = H_i K.$$

Proof. It is obvious that

$$y_i = (a_i, b_i)\widetilde{w}_i + \theta_i = (a_i, b_i)\overline{w}_i + a_i H_i G\theta + \theta_i = (a_i, b_i)\overline{w}_i + z_i,$$

where $z_i = a_i H_i G\theta + \theta_i$ is a zero-mean disturbance, *uncorrelated* with the elements of the 'expected' input vector $\overline{w}_i$. According to (6.14) we obtain that

$$\Delta(\Psi_{iN})\Delta^T(\Psi_{iN}) = \frac{1}{\sqrt{N}}\Theta_{iN}\frac{1}{\sqrt{N}}\Psi_{iN}^T.$$

$$\cdot \left(\frac{1}{N}\widetilde{W}_{iN}\Psi_{iN}^T\right)^{-1}\left(\frac{1}{N}\Psi_{iN}\widetilde{W}_{iN}^T\right)^{-1}.$$

$$\cdot \frac{1}{\sqrt{N}}\Psi_{iN}\frac{1}{\sqrt{N}}\Theta_{iN}^T,$$

and making use of the property that

$$\lambda_{\max}\left(\Delta(\Psi_{iN})\Delta^T(\Psi_{iN})\right) = \lambda_{\max}\left(\Delta^T(\Psi_{iN})\Delta(\Psi_{iN})\right),$$

for N large and $\Psi_{iN}^* = \overline{W}_{iN} = [\overline{w}_i^{(1)}, \overline{w}_i^{(2)}, ..., \overline{w}_i^{(N)}]$ we simply get

$$Q(\Psi_{iN}^*) = \lambda_{\max}\left(\frac{1}{N}\overline{W}_{iN}\overline{W}_{iN}^T\right)^{-1}\mathrm{var}\theta_i.$$

Under Lemma 6 in [65], for each Ψ_{iN} it holds that

$$Q(\Psi_{iN}^*) \leq Q(\Psi_{iN})$$

with probability 1.   ∎

Since the matrix K is unknown, the result (6.16) is not constructive, but gives the general concept of using the estimates of the noise-free interactions x_i. In real applications we can use the approximation of K obtained by the least square method and implement the recursive version of the algorithm

$$(\widehat{a}_i^{i.v.}, \widehat{b}_i^{i.v.})_{(k)} = (\widehat{a}_i^{i.v.}, \widehat{b}_i^{i.v.})_{(k-1)} +$$

$$+ \left[y_i^{(k)} - (\widehat{a}_i^{i.v.}, \widehat{b}_i^{i.v.})_{(k-1)}\widetilde{w}_i^{(k)}\right]\psi_i^{(k)^T}P_{i,k},$$

where

$$P_{i,k} = \frac{P_{i,k-1} - P_{i,k-1}\widetilde{w}_i^{(k)}\psi_i^{(k)^T}P_{i,k-1}}{1 + \psi_i^{(k)^T}P_{i,k-1}\widetilde{w}_i^{(k)}}.$$

6.5 Nonlinear Dynamic Components

The algorithm presented in Section 6.4 can be directly applied for the nets of nonlinear dynamic systems with the FIR Hammerstein-type components (see Fig. 6.2). Each component is then described by the following equation

$$y_{i,k} = \sum_{j=0}^{n_i} \gamma_j^{(i)} \mu_i(u_{i,k-j}) + \sum_{j=0}^{p_i} \lambda_j^{(i)} \eta_i(x_{i,k-j}) + z_{i,k}, \qquad (6.17)$$

where $\{\gamma_j^{(i)}\}_{j=0}^{n_i}$, and $\{\lambda_j^{(i)}\}_{j=0}^{p_i}$ are unknown impulse responses of the linear dynamic blocks in the ith component, with known orders n_i and p_i, respectively. Similarly, the functions $\mu_i()$ and $\eta_i()$ represents unknown nonlinearities in both channels in the ith component of the complex system of known parametric representation

$$\mu_i(u) = \sum_{t=1}^{m_i} c_t^{(i)} f_t^{(i)}(u), \qquad (6.18)$$

$$\eta_i(y) = \sum_{l=1}^{q_i} d_l^{(i)} g_l^{(i)}(y), \qquad (6.19)$$

with given basis functions $\{f_t^{(i)}(u)\}_{t=1}^{m_i}$ and $\{g_l^{(i)}(y)\}_{l=1}^{q_i}$. The signal $z_{i,k}$ is additive random zero-mean output noise. The goal is to estimate both the parameters $\{c_t^{(i)}\}_{t=1}^{m_i}$ and $\{d_l^{(i)}\}_{l=1}^{q_i}$ of nonlinearities $\mu_i(u)$ and $\eta_i(y)$, and the impulse responses $\{\gamma_j^{(i)}\}_{j=0}^{n_i}$, and $\{\lambda_j^{(i)}\}_{j=0}^{p_i}$ of the linear dynamic blocks, using input-output measurements of the whole complex system, presented in Fig. 6.1. Introducing the vectors of mixed products of parameters

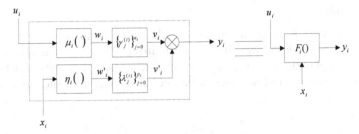

Fig. 6.2 Hammerstein model of the component

$$\theta^{(i)} = \left(\gamma_0^{(i)} c_1^{(i)}, ..., \gamma_0^{(i)} c_{m_i}^{(i)}, ..., \gamma_{n_i}^{(i)} c_1^{(i)}, ..., \gamma_{n_i}^{(i)} c_{m_i}^{(i)}\right)^T,$$

$$\vartheta^{(i)} = \left(\lambda_1^{(i)} d_1^{(i)}, ..., \lambda_1^{(i)} d_{q_i}^{(i)}, ..., \lambda_{p_i}^{(i)} d_1^{(i)}, ..., \lambda_{p_i}^{(i)} d_{q_i}^{(i)}\right)^T,$$

and the regressors

$$\phi_k^{(i)} = \left(f_1^{(i)}(u_{i,k}), ..., f_{m_i}^{(i)}(u_{i,k}), ..., f_1^{(i)}(u_{i,k-n_i}), ..., f_{m_i}^{(i)}(u_{i,k-n_i})\right)^T,$$

$$\varphi_k^{(i)} = \left(g_1^{(i)}(y_{i,k-1}), ..., g_{q_i}^{(i)}(y_{i,k-1}), ..., g_1^{(i)}(y_{i,k-p_i}), ..., g_{q_i}^{(i)}(y_{i,k-p_i})\right)^T,$$

we can show (6.17) in the compact form

$$y_{i,k} = \phi_k^{(i)T} \theta^{(i)} + \varphi_k^{(i)T} \vartheta^{(i)} + z_k.$$

The global matrices A and B (see (6.2) and (6.3)) and the input vectors are, for the Hammerstein components, defined as follows

$$A = \mathrm{diag}(\theta^{(1)}, \theta^{(2)}, ..., \theta^{(n)}),$$
$$B = \mathrm{diag}(\vartheta^{(1)}, \vartheta^{(2)}, ..., \vartheta^{(n)}),$$

and

$$\overline{u} = \left(\phi^{(1)T}, \phi^{(2)T}, ..., \phi^{(n)T}\right)^T,$$

$$\overline{x} = \left(\varphi^{(1)T}, \varphi^{(2)T}, ..., \varphi^{(n)T}\right)^T,$$

$$y = (y_1, y_2, ..., y_n)^T.$$

The measurement equation has the form

$$Y_{iN} = (\theta^{(i)}, \vartheta^{(i)}) W_{iN} + Z_{iN},$$

where

$$Y_{iN} = [y_{i,1}, y_{i,2}, ..., y_{i,N}], \qquad Z_{iN} = [z_{i,1}, z_{i,2}, ..., z_{i,N}],$$

$$W_{iN} = [w_{i,1}, w_{i,2}, ..., w_{i,N}], \qquad w_{i,k} = \left(\varphi_k^{(i)T}, \phi_k^{(i)T}\right)^T,$$

and the least squares and instrumental variables estimates

$$(\widehat{\theta^{(i)}}^{l.s.}, \widehat{\vartheta^{(i)}}^{l.s.}) = Y_{iN} W_{iN}^T \left(W_{iN} W_{iN}^T\right)^{-1},$$

$$(\widehat{\theta^{(i)}}^{i.v.}, \widehat{\vartheta^{(i)}}^{i.v.}) = Y_{iN} \Psi_{iN}^T \left(W_{iN} \Psi_{iN}^T\right)^{-1}$$

can be computed analogously as in Sections 6.3 and 6.4.

6.6   Simulation Example

In this section we present the performance of the algorithm on the example of
the simple, two-element linear static cascade system with feedback (see Fig.
6.3). We set $(a_1, b_1) = (1, 1)$, and $(a_2, b_2) = (2, 2)$ and the interconnections
are coded as follows

$$H = \begin{bmatrix} 0 & 1 \\ 1 & 0 \end{bmatrix}.$$

The system is excited by two independent uniformly random processes
$u_1, u_2 \sim U(0, 1)$, and disturbed by zero-mean noises $\delta_1, \xi_2 \sim U(-0.1, 0.1)$.
For both elements, the instrumental variables estimates are computed and
compared with the least squares results. In Fig. 6.4 we present the Euclidean
norms of the empirical error Δ for both algorithms.

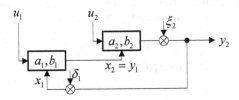

Fig. 6.3 The two-element cascade system with feedback

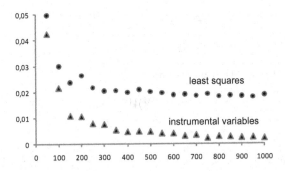

Fig. 6.4 The estimation error Δ versus number of measurements N

6.7   Final Remarks

The idea of instrumental variables estimate was successfully generalized for
the complex, interconnected systems. Since the production and transporta-
tion systems usually work in steady state, we limited ourselves to the static
blocks. However, generalization for the FIR linear dynamic components seems
to be quite simple. In the contrary to the traditional least squares approach,

the proposed algorithm recovers true parameters of subsystems. General conditions are imposed on the instrumental variables for to estimate to be consistent, and then, the form of optimal values of instruments is shown. We emphasize that the method works for any structure of the system and for any distribution and correlation of the random noises.

The algorithms presented in the chapter are of key importance in problems of optimal control in large-scale interconnected systems. Under numerous decision variables and parameters, global optimization of the control quality index is often excluded by reason of high computational complexity. Then, the only possible choice is to divide global problem into subproblems, which can be optimize in parallel. Obviously, good models of particular blocks are needed in local optimization tasks. Moreover, by virtue of constraints imposed on inputs, local problems must be coordinated to assure admissibility of the solution. We refer the reader to [32] for details concerning the concept of decomposition and coordination in hierarchical systems.

Chapter 7
Structure Detection and Model Order Selection

The chapter deals with the problem of automatic model selection of the nonlinear characteristic in a block-oriented dynamic system (see [68]). We look for the best parametric model of Hammerstein system nonlinearity. The finite set of candidate classes of parametric models is given and we select one of them on the basis of the input-output measurement data, using the concept of nearest neighbour borrowed from pattern recognition techniques. The algorithm uses the pattern of the true characteristic generated by its nonparametric estimates on the grid of fixed (e.g. equidistant) points. Each class generates parametrized learning sequence through the values on the same grid of points. Next, for each class, the optimal parameters are computed by the least squares method. Finally, the nearest neighbour approach is applied for the selection of the best model in the mean square sense.

The idea is illustrated by the example of competition between polynomial, exponential and piece-wise linear models of the same complexity (i.e., number of parameters needed to be stored in memory). For all classes, the upper bounds of the integrated approximation errors of the true characteristic are computed and compared.

7.1 Introduction

7.1.1 Types of Knowledge. Classification of Approaches

The procedure of system identification is always preceded by making decision about the assumed structure of model, e.g. kind of basis functions of the nonlinear characteristic. Although the resulting model is very sensitive to prior assumptions, this issue is treated very cursory/rarely in the literature (see e.g. [1], [11], and [54]). Obviously, if the assumed model is not correct, then the estimate has systematic approximation error, which does not tend to zero even for large number of data.

G. Mzyk, *Combined Parametric-Nonparametric Identification of Block-Oriented Systems*, Lecture Notes in Control and Information Sciences 454, DOI: 10.1007/978-3-319-03596-3_7, © Springer International Publishing Switzerland 2014

In the chapter we propose the method of recognition of the best model from the given set of candidates. In particular, we consider competition between polynomial, exponential and piecewise linear models of the same complexity, i.e., we select the class which guarantees the best accuracy with the same number of parameters needed to be stored in memory.

We propose the cooperation between parametric and nonparametric methods for system modeling along with nearest neighbour pattern recognition technique. The term 'parametric' means that estimated nonlinearity will be approximated (see e.g. [27] and [28]) with the use of the model consisting of *finite* and *known* number of parameters. The nonparametric approach plays the role of the data pre-filtering with compression and is applied here to support smart selection of the best parametric model of nonlinear characteristic from a given finite number of possible types of models. In Section 7.2 the problem is formulated in detail and the close relation between the nonparametric regression function and the static characteristic of Hammerstein system is discussed. We present the nonparametric method of regression function estimation, and next, in Section 7.3, the traditional parametric nonlinear least-squares for the regression function approximation in Hammerstein system by various kinds of parametric models is reminded. The above two different approaches are thus combined. First, nonparametric estimates are pointwisely computed from the learning pairs on the grid of N_0 input points, and next, they support selection of one from competing parametric models. The parameters of the best model in the considered family of various models are then obtained by the nonlinear least squares, and broad variety of optimization algorithms (also soft methods, e.g., genetic, tabu search, simulated annealing, particle swarming) can be applied in this stage, depending on the specifics of the optimization criterion. The best model is then indicated as the nearest neighbour of the pattern of the system, established by the non-parametric estimate in stage 1. If the resulting parametric approximation is not satisfying, and the number of measurements is large enough, the residuum between the system output and the model output can be used for nonparametric refinement of the model [126].

7.1.2 Contribution

The novelty of the approach consists in the following:

- the idea of the *regression-based* approach to *parametric* approximation of nonlinear characteristic in Hammerstein system is introduced,
- the proposed identification method works under *lack* or *uncertain* knowledge about the form of the static nonlinear characteristic,
- models which are *not linear in the parameters* are admitted,
- the strategy allows to refine the model, when the obtained approximation, resulting the parametric approximation, is not satisfying, and the data-set is long enough to apply the purely nonparametric correction.

7.2 Statement of the Problem

7.2.1 The System

The Hammerstein system is considered, i.e., a static non-linearity $\mu()$, and a linear dynamics, with the impulse response $\{\gamma_i\}_{i=0}^{\infty}$, connected in a cascade and described by the following set of equations: $y_k = v_k + z_k$, $v_k = \sum_{i=0}^{\infty} \gamma_i w_{k-i}$, $w_k = \mu(u_k)$, or equivalently

$$y_k = \sum_{i=0}^{\infty} \gamma_i \mu(u_{k-i}) + z_k, \tag{7.1}$$

where u_k and y_k denote the system input and output at time k, respectively, and z_k is the output noise (see Fig. 7.1)

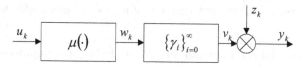

Fig. 7.1 The identified Hammerstein system

7.2.2 Assumptions

We assume that:

Assumption 7.1. *The input signal $\{u_k\}$ is for $k = ..., -1, 0, 1, ...$ an i.i.d. bounded random process, with strictly positive probability density function $\nu(u)$ on the interval (u_L, u_U), i.e., $0 < \nu_{\min} \leq \nu(u) \leq \nu_{\max} < \infty$ for $u_L < u < u_U$, some $u_L, u_U < \infty$, and $\nu(u) = 0$ elsewhere. We also denote*

$$u_{\max} = \max(|u_L|, |u_U|).$$

Assumption 7.2. *The nonlinear characteristic $\mu(u)$ is a bounded function on the interval $[u_L, u_U]$, i.e.,*

$$|\mu(u)| \leqslant w_{\max}, \tag{7.2}$$

where $w_{\max}$ is some positive constant, but is completely unknown.

Assumption 7.3. *The linear dynamics is an asymptotically stable IIR filter*

$$v_k = \sum_{i=0}^{\infty} \gamma_i w_{k-i}, \tag{7.3}$$

with the unknown impulse response $\{\gamma_i\}_{i=0}^{\infty}$ (i.e., such that $\sum_{i=0}^{\infty} |\gamma_i| < \infty$).

Assumption 7.4. *The output noise $\{z_k\}$ is a random, arbitrarily correlated linear process, governed by the general equation*

$$z_k = \sum_{i=0}^{\infty} \omega_i \varepsilon_{k-i}, \qquad (7.4)$$

where $\{\varepsilon_k\}$, $k = ..., -1, 0, 1, ...$, is a bounded stationary zero-mean white noise ($E\varepsilon_k = 0$, $|\varepsilon_k| \leqslant \varepsilon_{max}$), independent of the input signal $\{u_k\}$, and $\{\omega_i\}_{i=0}^{\infty}$ is unknown; $\sum_{i=0}^{\infty} |\omega_i| < \infty$. Hence, the noise $\{z_k\}$ is also a stationary zero-mean and bounded process $|z_k| \leqslant z_{max}$, where $z_{max} = \varepsilon_{max} \sum_{i=0}^{\infty} |\omega_i|$.

Assumption 7.5. *For clarity let $\mu(u_0)$ is known at some point u_0 and let $\gamma_0 = 1$.*

Assumption 7.6. *Only u_k and y_k, i.e., external system signals, can be measured, and in particular the internal signal w_k is not admissible for a direct measurements.*

As it was explained in detail in [64] and [65], the input-output pair $(u_0, \mu(u_0))$ assumed to be known can refer to arbitrary $u_0 \in [-u_{max}, u_{max}]$, and hence we shall further assume for convenience that $u_0 = 0$ and $\mu(0) = 0$, without loss of generality.

7.2.3 Preliminaries

The renowned dependence between the regression function of the output on the input of the system and its nonlinear characteristic

$$R(u) = E\{y_k | u_k = u\} = \gamma_0 \mu(u) + d, \qquad (7.5)$$

where $d = E\mu(u_k) \cdot \sum_{i>0} \gamma_i = \text{const}$. Since by assumption $\mu(0) = 0$, it holds that $R(0) = d$ and

$$R(u) - R(0) = \gamma_0 \mu(u) \qquad (7.6)$$

which, along with $\gamma_0 = 1$ in Assumption 7.5 yields

$$\mu(u) = R(u) - R(0). \qquad (7.7)$$

Remark 7.1. *The equivalence (7.6) allows to recover the nonlinear characteristic $\mu()$ under lack of prior knowledge about the linear dynamic subsystem. This observation was successfully utilized in eighties by Greblicki and Pawlak for construction of nonparametric identification methods, when the form of $\mu()$ is also unknown, but was never explored in parametric identification framework, when some (even uncertain) prior knowledge of only $\mu()$ is given.*

Below, we show that the regression-based approach to nonlinearity recovering in Hammerstein system, originated from the relation (7.7), can also be applied when some parametric model, or a set of such models due to uncertainty of the form of characteristic, is taken into account.

7.3 Parametric Approximation of the Regression Function

The nonparametric identification methods recover the true regression function (7.5) 'point by point'. In this section we admit the parametric model of nonlinear static characteristic $\mu()$ and show how to find the best approximation of the characteristic in the form of the assumed model directly from the input-output measurements (u_k, y_k) (despite the fact that the signal $w_k \stackrel{\bullet}{=} \mu(u_k)$, cannot be measured).

7.3.1   The Best Model

We thus suspect that the nonlinear characteristic $\mu(u)$ can be well approximated by the model from the given class

$$\bar{\mu}(u, c), \tag{7.8}$$

where $c = (c_1, c_2, ..., c_m)^T$ embraces finite number of parameters. The function $\bar{\mu}(u, c)$ is further assumed to be *differentiable* with respect to c. Let $c^* = (c_1^*, c_2^*, ..., c_m^*)^T$ be the best (unique) choice of c in the sense that

$$c^* = \arg\min_c E\left(\mu(u) - \bar{\mu}(u, c)\right)^2. \tag{7.9}$$

Further, owing to (7.5), we will explore the generalized version of the class (7.8)

$$\bar{\mu}(u, \vartheta) = c_\alpha \bar{\mu}(u, c) + c_\beta \tag{7.10}$$

where $\vartheta = \left(c^T, c_\alpha, c_\beta\right)^T$ is the extended model vector enriched with the scale c_α and the offset c_β. Obviously for $c_\alpha = 1$ and $c_\beta = 0$ (see (7.7)) equation (7.9) can be rewritten in the form

$$\arg\min_\vartheta E\left(\mu(u) - \bar{\mu}(u, \vartheta)\right)^2 = \left(c^{*T}, 1, 0\right)^T \triangleq \vartheta^*. \tag{7.11}$$

Remark 7.2. *For the models $\bar{\mu}(u, c)$, which are linear in the parameters and have additive constant (e.g. polynomials) $\bar{\mu}(u, c)$ and $\bar{\mu}(u, \vartheta)$ are indistinguishable. For example the polynomial model of order $m - 1$*

$$\bar{\mu}(u, c) = c_m u^{m-1} + ... + c_2 u + c_1$$

is equivalent to

$$\overline{\mu}(u, \vartheta) = c_\alpha c_m u^{m-1} + \ldots + c_\alpha c_2 u + c_\alpha c_1 + c_\beta.$$

In some special cases, the minimization criterion (7.11) has not unique solution (see Remarks 7.3 and 7.4 below).

Remark 7.3. *If the model $\overline{\mu}(u, c)$ is linear with respect to parameters, i.e.,*

$$\overline{\mu}(u, c) = \sum_{l=1}^{m} c_l f_l(u) = \phi^T(u)c, \qquad (7.12)$$

where $\phi(u) = (f_1(u), f_2(u), \ldots, f_m(u))^T$ includes the set of known basis functions, then

$$E\left(\mu(u) - \overline{\mu}(u, \vartheta)\right)^2$$

is minimized by any vector of the form

$$\vartheta_a = \left(ac^{*T}, \frac{1}{a}, 0\right)^T, \qquad a \neq 0.$$

Remark 7.4. *If the model $\overline{\mu}(u, c)$ has additive parameter c_{l_0}, i.e., $\frac{\partial \overline{\mu}(u,c)}{\partial c_{l_0}} = 1$, then $E\left(\mu(u) - \overline{\mu}(u, \vartheta)\right)^2$ is minimized by each vector of the form*

$$\vartheta_b = (c_1^*, \ldots, c_{l_0-1}, c_{l_0} + b, c_{l_0+1}, \ldots, c_m^*, 1, -b)^T, \qquad b \in R.$$

Since $R(u) = \gamma_0 \mu(u) + d$ (see (7.5)), let $\overline{R}(u, \theta) = p_1 \overline{\mu}(u, c) + p_2$, where $\theta = (c^T, p_1, p_2)^T$, be respective class of approximation models of $R(u)$. The function $\overline{R}(u, \theta)$ is obviously differentiable with respect to θ.

Remark 7.5. *If $E\left(\mu(u) - \overline{\mu}(u, \vartheta)\right)^2$ is minimized by $\left(c^{*T}, 1, 0\right)^T$, then*

$$E\left(R(u) - \overline{R}(u, \theta)\right)^2$$

is minimized by $\left(c^{*T}, \gamma_0, d\right)^T$.

7.3.2 Approximation of the Regression Function

By rewriting (7.1) in the form

$$y_k = \gamma_0 \mu(u_k) + \sum_{i=1}^{\infty} \gamma_i \mu(u_{k-i}) + z_k,$$

and taking into account that

$$R(u_k) = \gamma_0 \mu(u_k) + \sum_{i=1}^{\infty} \gamma_i E\mu(u_1),$$

we obtain the equivalent (cardinal) description

$$y_k = R(u_k) + \delta_k \qquad (7.13)$$

of the Hammerstein system (see Fig. 7.2), in which the total noise

$$\delta_k \triangleq y_k - R(u_k) = \sum_{i=1}^{\infty} \gamma_i \left(\mu(u_{k-i}) - E\mu(u_1)\right) + z_k$$

is zero-mean ($E\delta_k = 0$) and independent of u_k. As it was noticed before, we want to find the vector θ^* for which the model $\overline{R}(u_k, \theta)$ fits to data the best, in the sense of the following criterion

$$E\left(y_k - \overline{R}(u_k, \theta)\right)^2 =$$
$$= E\left[y_k - R(u_k) + R(u_k) - \overline{R}(u_k, \theta)\right]^2$$
$$= E\left[(y_k - R(u_k))^2 + 2\left(y_k - R(u_k)\right)\left(R(u_k) - \overline{R}(u_k, \theta)\right) + \left(R(u_k) - \overline{R}(u_k, \theta)\right)^2\right]$$
$$= E\left(y_k - R(u_k)\right)^2 + 2E\left[(y_k - R(u_k))\left(R(u_k) - \overline{R}(u_k, \theta)\right)\right] + E\left(R(u_k) - \overline{R}(u_k, \theta)\right)^2$$
$$= \mathrm{var}\delta_k + 2E\left[\delta_k\left(R(u_k) - \overline{R}(u_k, \theta)\right)\right] + E\left(R(u_k) - \overline{R}(u_k, \theta)\right)^2 =$$
$$= \mathrm{var}\delta_k + 0 + E\left(R(u_k) - \overline{R}(u_k, \theta)\right)^2 \qquad (7.14)$$

From (7.14) we conclude that

$$\arg\min_{\theta} E\left(y_k - \overline{R}(u_k, \theta)\right)^2 = \arg\min_{\theta} E\left(R(u_k) - \overline{R}(u_k, \theta)\right)^2 = \theta^*, \qquad (7.15)$$

which is fundamental for the empirical least squares approximation, presented in the next section.

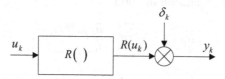

Fig. 7.2 Equivalent (regression-oriented) description of Hammertein system

7.3.3 The Nonlinear Least Squares Method

Due to the traditional approach, the model would be computed on the basis of large number N of measurements $\{(u_k, y_k)\}_{k=1}^N$ with noisy y_k (see (7.13) and Fig. 7.2). The parameters would be obtained by minimization of the following empirical criterion

$$\widehat{Q}(\theta) = \frac{1}{N} \sum_{k=1}^N \left(y_k - \overline{R}(u_k, \theta)\right)^2 \tag{7.16}$$

with respect to θ, which would lead to the parameter vector estimate

$$\widehat{\theta}_N = \arg\min_\theta \widehat{Q}(\theta) \tag{7.17}$$

of the best approximation. Such a procedure is very badly conditioned and complicated numerically for the two reasons:

- in general, the optimization task is nonlinear, and the only choice is to apply the iterative searching method of the optimum (gradient method, Newton method, Levenberg-Marquardt method, genetic algorithms, simulated annealing, tabu search, etc.),
- computing of the value of the cost function $\widehat{Q}(\theta)$ for a given θ (see (7.16)), in each individual iteration, requires N operations.

Therefore, instead of N input-output pairs $\{(u_k, y_k)\}_{k=1}^N$ we propose to use much shorter sequence (of the length $N_0 << N$) of the pre-filtered (denoised and compressed) data $\{(\overline{u}_i, \widehat{R}_N(\overline{u}_i))\}_{i=1}^{N_0}$, where $\overline{u}_i$'s are freely selected points, e.g., uniformly distributed on the interval $(-u_L, u_U)$, and $\widehat{R}_N(\overline{u}_i)$'s are the nonparametric estimates of the regression function values in this points, e.g., computed by the kernel estimation method ([50])

$$\widehat{R}_N(\overline{u}_i) = \frac{\sum_{k=1}^N y_k K(\frac{u_k - \overline{u}_i}{h})}{\sum_{k=1}^N K(\frac{u_k - \overline{u}_i}{h})}, \tag{7.18}$$

or by the orthogonal expansion method

$$\widehat{R}_N(\overline{u}_i) = \frac{\sum_{r=1}^q \widehat{a}_r \varphi_r(\overline{u}_i)}{\sum_{r=1}^q \widehat{b}_r \varphi_r(\overline{u}_i)}, \tag{7.19}$$

where $\widehat{a}_r = \frac{1}{N} \sum_{k=1}^N y_k \varphi_r(u_k)$, $\widehat{b}_r = \frac{1}{N} \sum_{k=1}^N \varphi_r(u_k)$, $i = 1, 2, ..., N_0$, $K()$ is a kernel function, $h = h(N)$ – a bandwidth parameter, $q = q(N)$ – a cut-off level (scale), $\{\varphi_r()\}$ – an orthonormal set of basis functions, and $\widehat{a}_r$, $\widehat{b}_r$ are the coefficients, obtained by orthogonal projection.

7.4 Model Quality Evaluation

Assume that the true regression function $R(u)$ fulfils the Lipschitz condition, i.e., for each $u, x \in (u_L, u_U)$ it holds that

$$|R(u) - R(x)| \leq L_R |u - x|, \qquad (7.20)$$

where $L_R < \infty$ is known constant. Since the cumulative distribution function of u, i.e. $F(u) = \int_{u_L}^{u} \nu(u) du$ is continuous and invertible (see Assumption 7.1), let us define the following points (nodes) of nonlinearity approximation

$$\overline{u}_i = F^{-1}\left(\frac{i}{N_0}\right), \qquad (7.21)$$

and from the inequality $\nu(u) \geq \nu_{\min}$ we note that

$$F(\overline{u}_{i+1}) - F(\overline{u}_i) \geq \nu_{\min}(\overline{u}_{i+1} - \overline{u}_i).$$

Since

$$F(\overline{u}_{i+1}) - F(\overline{u}_i) = \frac{1}{N_0},$$

we obtain that

$$\overline{u}_{i+1} - \overline{u}_i \leq \frac{1}{N_0 \nu_{\min}}.$$

For the N_0 fixed points $\{\overline{u}_i\}_{i=1}^{N_0}$, let us denote by $\overline{R}_\Lambda(u)$ the piecewise linear model of $R(u)$ spanned on the nodes $\{(\overline{u}_i, R(\overline{u}_i))\}_{i \in \Lambda}$, where $\Lambda = \{1, 2, ..., N_0\}$. For the random input u, let us also define the mean squared error $E\left(\overline{R}_\Lambda(u) - R(u)\right)^2$, which can be upper-bounded as follows

$$E\left(\overline{R}_\Lambda(u) - R(u)\right)^2 = \int_{-\infty}^{\infty} \left(\overline{R}_\Lambda(u) - R(u)\right)^2 \nu(u) du =$$

$$= \int_{u_L}^{u_U} \left(\overline{R}_\Lambda(u) - R(u)\right)^2 \nu(u) du =$$

$$= \sum_{i=1}^{N_0-1} \int_{\overline{u}_i}^{\overline{u}_{i+1}} \left(\overline{R}_\Lambda(u) - R(u)\right)^2 \nu(u) du \leq$$

$$\leq N_0 \frac{1}{N_0 \nu_{\min}} \left(\frac{1}{N_0 \nu_{\min}} L_R\right)^2 \nu_{\max}.$$

Hence

$$E\left(\overline{R}_\Lambda(u) - R(u)\right)^2 \leq \Delta_0,$$

where

$$\Delta_0 = \frac{L_R^2 \nu_{\max}}{N_0^2 \nu_{\min}^3} \sim \mathcal{O}\left(\frac{1}{N_0^2}\right) \qquad (7.22)$$

can be made arbitrarily small by proper selection of N_0. Analogously, we evaluate the quality of each of l-th candidates for the parametric model

$$E \left(\overline{R}^{(l)} (u, \theta_l) - R(u) \right)^2 \le$$

$$\le 2E \left(\overline{R}^{(l)} (u, \theta_l) - \overline{R}_\Lambda(u) \right)^2 + 2E \left(\overline{R}_\Lambda(u) - R(u) \right)^2 \le$$

$$\le 2\nu_{\max} \int_{u_L}^{u_U} \left(\overline{R}^{(l)} (u, \theta_l) - \overline{R}_\Lambda(u) \right)^2 du + 2\Delta_0, \qquad (7.23)$$

where

$$\int_{u_L}^{u_U} \left(\overline{R}^{(l)} (u, \theta_l) - \overline{R}_\Lambda(u) \right)^2 du \le \qquad (7.24)$$

$$\le \frac{1}{N_0 \nu_{\min}} \sum_{i=1}^{N_0} \left(\left| \overline{R}^{(l)} (\overline{u}_i, \theta_l) - \overline{R}_\Lambda(\overline{u}_i) \right| + \frac{L}{N_0 \nu_{\min}} \right)^2 = E_l(\theta_l),$$

$L = L_R + L_l$, and L_l is the Lipschitz constant of the l-th model, assumed to be possible for computation. Since it holds that

$$E \left(\overline{R}^{(l)} (u, \theta_l) - R(u) \right)^2 \le \Delta_l(\theta_l),$$

where

$$\Delta_l(\theta_l) = 2\nu_{\max} E_l(\theta_l) + \Delta_0, \qquad (7.25)$$

the quality of the model will be further verified and compared with the use of guaranteed accuracy Δ_l, which can be simply computed, for each class and given θ_l's, on the basis of the estimated pattern $\left\{ \widehat{R}_N(\overline{u}_i) \right\}_{i=1}^{N_0}$ of the true characteristic, the parametrized pattern $\overline{R}^{(l)} (\overline{u}_i, \theta_l)$ of the l-th class, and a priori knowledge of $\nu_{\min}, \nu_{\max}, L_R$, and L_l.

7.5 The Algorithm of the Best Model Selection

In this section we present the scheme of the method for the best model recognition (see Fig. 7.3). In general, we are given M parametric classes of model

$$\left\{ \overline{R}^{(1)} (u, \theta_1) \right\}, \left\{ \overline{R}^{(2)} (u, \theta_2) \right\}, ..., \left\{ \overline{R}^{(M)} (u, \theta_M) \right\}, \qquad (7.26)$$

[and optionally the weights (costs) $\varkappa_1, \varkappa_2, ..., \varkappa_M$, representing priorities of these models (e.g. dependent on the computational complexity, properties, or requirements).] The proposed strategy consists of the following three steps.

Step 1. Nonparametric smoothing (denoising and data-compression)
For the grid of N_0 selected fixed points $\overline{u}_1, \overline{u}_2, ..., \overline{u}_{N_0}$, (see (7.21)) compute nonparametric estimates $\widehat{R}_N(\overline{u}_1), \widehat{R}_N(\overline{u}_2), ..., \widehat{R}_N(\overline{u}_{N_0})$ of the regression function (7.5), playing the role of the true characteristic $\mu()$.

Step 2. Least squares approximation
For each class $l = 1, 2, ..., M$ minimize the loss function

$$\widehat{Q}_l(\theta_l) = \frac{1}{N_0} \sum_{i=1}^{N_0} \left(\widehat{R}_N(\overline{u}_i) - \overline{R}^{(l)}(\overline{u}_i, \theta_l) \right)^2, \qquad (7.27)$$

being the empirical counterpart of the least squares criterion

$$Q_l(\theta_l) = \frac{1}{N_0} \sum_{i=1}^{N_0} \left(R(\overline{u}_i) - \overline{R}^{(l)}(\overline{u}_i, \theta_l) \right)^2,$$

with respect to θ_l, getting the estimate

$$\widehat{\theta}_l^* = \arg\min_{\theta_l} \widehat{Q}_l(\theta_l) \qquad (7.28)$$

of the best parameters θ_l^* in the l-th class.

Step 3. Nearest neighbour model selection
Select the 'nearest neighbour' model, i.e.,

$$\left\{ \overline{R}^{(l_0)}\left(u, \widehat{\theta}_{l_0}^*\right) \right\}, \qquad (7.29)$$

where

$$l_0 = \arg\min_l [\varkappa_l] \Delta_l(\widehat{\theta}_l^*). \qquad (7.30)$$

Remark 7.6. *Since N_0 is typically much less than the number of measurements N, and the nonparametric estimates $\left\{ \widehat{R}_N(\overline{u}_i) \right\}_{i=1}^{N_0}$ are computed only once, i.e., they are common for each tested models, the algorithm of model selection works very fast. Nevertheless, if the guaranteed evaluation of the distance between the model and the pattern, proposed in (7.25), takes into consideration only moderate number of operating points $\{\overline{u}_i\}_{i=1}^{N_0}$, the component Δ_0 (see (7.22)) cannot be neglected. It means that selection of N_0 and distribution of $\overline{u}_i$'s are of high importance here.*

Above strategy is tested numerically in Section 7.8 on the example of polynomial, exponential or piecewise-linear model recognition.

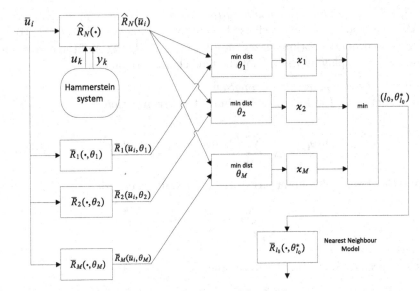

Fig. 7.3 Scheme of the 3-step method of model selection

7.6 Limit Properties

The following theorem holds.

Theorem 7.1. *Let*

$$l^* = \arg\min_l E\left(\overline{R}^{(l)}(u,\theta_l^*) - R(u)\right)^2, \tag{7.31}$$

be the true index of the best approximation class in the sense of integrated squared error, and θ_l^*'s are, for $l = 1,2,...,M$, defined by (7.15) and let $\overline{R}^{(l)}(u,\theta_l)$ be continuous and differentiable with respect to parameters θ_l, i.e.,

$$\left\|\nabla_\theta\left(\overline{R}^{(l)}(u,\theta_l)\right)\right\| \le D_{l,\max} < \infty \tag{7.32}$$

in the set of admissible parameters $\theta_l \in \Theta$. If $N_0, N \to \infty$, then

$$l_0 \to l^* \text{ in probability.}$$

Proof. From (7.22) we have that $\lim_{N_0 \to \infty} \Delta_0 = 0$. Thus, the equation (7.25) asymptotically takes the form $\Delta_l(\theta_l) = 2\nu_{\max}E_l(\theta_l)$ with probability 1. It leads to the following equivalence

$$l_0 = \arg\min_l \Delta_l(\widehat{\theta}_l^*) = \arg\min_l E_l(\widehat{\theta}_l^*).$$

For $N_0 \to \infty$ we have

$$E_l(\theta_l) = \frac{1}{N_0 \gamma_{\min}} \sum_{i=1}^{N_0} \left(\overline{R}^{(l)}(\overline{u}_i, \theta_l) - R(\overline{u}_i) \right)^2 = \frac{1}{\gamma_{\min}} Q_l(\theta_l)$$

and from (7.23) we get

$$E \left(\overline{R}^{(l)}(u, \theta_l) - R(u) \right)^2 \leq 2\gamma_{\max} E_l(\theta_l) = c \cdot \frac{1}{N_0} \sum_{i=1}^{N_0} \left(\overline{R}^{(l)}(\overline{u}_i, \theta_l) - R(\overline{u}_i) \right)^2$$

where $c = \frac{2\gamma_{\max}}{\gamma_{\min}}$. Under consistency of nonparametric estimates (7.18) and (7.19) it holds that

$$\widehat{Q}_l(\theta_l) \rightarrow \frac{1}{N_0} \sum_{i=1}^{N_0} \left(\overline{R}^{(l)}(\overline{u}_i, \theta_l) - R(\overline{u}_i) \right)^2 = \gamma_{\min} E_l(\theta_l) = Q_l(\theta_l)$$

in probability as $N \rightarrow \infty$. From (7.32) also the estimate (7.28) converges to the parameters of the best approximation, i.e.,

$$\widehat{\theta}_l \rightarrow \theta_l^*$$

in probability (see the proof of Theorem 1 in [65]). Consequently, by continuity of the cost function $Q_l(\theta_l)$ with respect to θ_l it also holds that

$$Q_l(\widehat{\theta}_l) \rightarrow Q_l(\theta_l^*).$$

Thus, for each pair of classes l_1 and l_2 it holds that

$$Q_{l_1}(\theta_{l_1}) < Q_{l_2}(\theta_{l_2}) \Leftrightarrow Q_{l_1}(\widehat{\theta}_{l_1}) < Q_{l_2}(\widehat{\theta}_{l_2})$$

with probability 1, as $N_0, N \rightarrow \infty$. ∎

Remark 7.7. *Under Assumption 7.1 the method of selection of the estimation points $\{\overline{u}_i\}_{i=1}^{N_0}$ given by (7.21) can be generalized for e.g. equidistant grid of points*

$$\overline{u}_i = u_L + \frac{i}{N_0}(u_U - u_L).$$

Such a choice obviously changes the constants in evaluations (7.22) and (7.24), but the proof of Theorem 7.1 remains valid.

Remark 7.8. *If one of the candidate classes represents the true system characteristic, then*

$$E_l(\theta_l^*) = 0$$

and, asymptotically, the algorithm selects this class with probability 1.

7.7　Computational Complexity

Assume we are given a dictionary of n basis functions and we want to select s of them. We obtain large number of candidate classes, i.e., $M = \binom{n}{s}$. Since the optimization (7.17) must be performed in all M classes, and computing of the criterion function (7.16) requires $O(N)$ operations, selection of the best model in a traditional way has the complexity not simpler that $O(MN)$. In the proposed approach N measurements are firstly compressed to moderate number of N_0 data points, with the complexity $O(N_0 N)$. Next, the least squares criterion is minimized in all classes with the use of the same N_0 points. For example, if $n = 10$, $s = 3$, $N = 100$ and $N_0 = 3$ we need ~ 12000 operations in the traditional parametric method, and only ~ 660 operations in the proposed procedure with nonparametric pre-filtering.

7.8　Examples

7.8.1　Example I – Voltage-Intensity Diode Characteristic

The strategy was tested on the simulated data, representing voltage-intensity relationship of the real diode (see e.g. [26]) in the presence of the random disturbances. The Hammerstein model was excited by the uniformly distributed i.i.d. random process $u_k \sim U(0,1)$, i.e., $u_{\max} = 1$. The true nonlinear characteristic was set as follows

$$\mu(u) = c_1^* \left(e^{c_2^* u} - 1 \right), \text{ with } c_1^* = c_2^* = 1, \tag{7.33}$$

i.e. $w_k = e^{u_k} - 1$. The nonlinear static block was followed by the linear dynamics with the infinite impulse response $\gamma_i = 2^{-i}$, realized by the following differential equation

$$v_k = \frac{1}{2} v_{k-1} + w_k. \tag{7.34}$$

The output v_k was disturbed by zero-mean colored random noise $z_k = 0.9 z_{k-1} + \varepsilon_k$, with $\varepsilon_k \sim U(-0.1, 0.1)$. The following three parametric classes of models were considered:

- exponential

$$\overline{R}^{(1)}(u, \theta_1) = \theta_{1,1} \left(e^{\theta_{1,2} u} - 1 \right), \qquad \theta_1 = (\theta_{1,1}, \theta_{1,2})^T,$$

- polynomial

$$\overline{R}^{(2)}(u, \theta_2) = \theta_{2,1} u^2 + \theta_{2,2} u = u \left(\theta_{2,1} u + \theta_{2,2} \right), \qquad \theta_2 = (\theta_{2,1}, \theta_{2,2})^T,$$

- piecewise linear

$$\overline{R}^{(3)}(u, \theta_3) = \begin{cases} \theta_{3,1}(u - \theta_{3,2}), \text{ as } u \geq \theta_{3,2} \\ 0, \text{ elsewhere} \end{cases}, \qquad \theta_3 = (\theta_{3,1}, \theta_{3,2})^T.$$

Remark 7.9. *Let us emphasize that the models $\overline{R}^{(1)}()$ and $\overline{R}^{(3)}()$ are not linear with respect to the parameters in θ_1, and θ_3. In consequence, the optimization task in* Step 2 *(see (7.28))*

$$Q_l(\theta_l) = \frac{1}{N_0} \sum_{i=1}^{N_0} \left(\widehat{R}_N(\overline{u}_i) - \overline{R}^{(l)}(\overline{u}_i, \theta_l) \right)^2 \to \min_{\theta_l}, \qquad l = 1, 3 \qquad (7.35)$$

is much more difficult that for the case of $\overline{R}^{(2)}()$, in which the standard least squares procedure can be applied, i.e.,

$$\widehat{\theta}_2 = \left(\Phi_{N_0}^T \Phi_{N_0} \right)^{-1} \Phi_{N_0}^T Y_{N_0,N},$$

where

$$\Phi_{N_0} = \begin{bmatrix} \overline{u}_1^2 & \overline{u}_1 \\ \overline{u}_2^2 & \overline{u}_2 \\ \vdots & \vdots \\ \overline{u}_{N_0}^2 & \overline{u}_{N_0} \end{bmatrix}, \text{ and } Y_{N_0,N} = \begin{bmatrix} \widehat{R}_N(\overline{u}_1) \\ \widehat{R}_N(\overline{u}_2) \\ \vdots \\ \widehat{R}_N(\overline{u}_{N_0}) \end{bmatrix}.$$

For (7.35), in the computer experiment we applied the standard *Matlab* procedures. The results are presented in Fig. 7.4 and in Fig. 7.5.

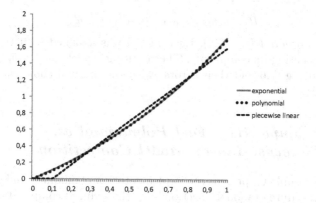

Fig. 7.4 Results of approximation of the diode characteristic for various models

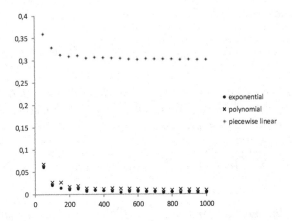

Fig. 7.5 Approximation error for various models versus number of measurements

7.8.2   Example II – Model Order Selection

To illustrate the applicability and usability of the method for supporting of model-order selection, we simulated the following Hammerstein system:

$$\mu(u) = \text{sgn}(u) + |u|, \quad v_k = 0.9v_{k-1} + w_k. \tag{7.36}$$

The random input u_k and random noise z_k were uniformly distributed on the intervals $[-1, 1]$ and $[-0.1, 0.1]$, respectively. In the kernel-type estimation algorithm (7.18) we applied the window kernel $K(x) = \begin{cases} 1, \text{ as } |x| < 1 \\ 0, \text{ elsewhere} \end{cases}$, and set $h(N) \sim N^{-1/5}$. We implemented competition between polynomial models

$$\overline{R}^{(l)}(u, \theta_l) = p_0 + p_1 u + \dots + p_l u^l, \tag{7.37}$$

with various orders l (see [95]). Each model was assigned with the penalty $\varkappa_l = l$ representing its complexity. The resulting models for $l = 0, 1, ..., 6$ are presented in Fig. 7.6 and their errors (with the optimal choice $l_0 = 3$) are shown in Fig. 7.7.

7.8.3   Example III – Fast Polynomial vs. Piecewise-Linear Model Competition

In the last example we present the implementation of modified and simplified heuristic procedure of model selection with the reduced computational complexity. The piecewise-linear model spanned on all nodes of nonparametric regression is reduced step-by-step by rejection of individual points and recalculating of the cost function. When the postulated accuracy is achieved, the obtained model is compared with the polynomial one, with the same number of parameters.

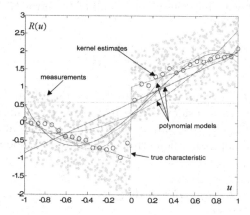

Fig. 7.6 Results of identification for various models

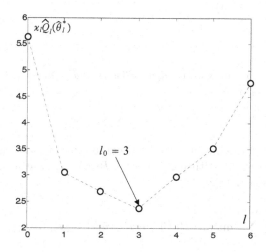

Fig. 7.7 The cost function versus the order of polynomial model

Efficient Piecewise Linear Approximation

For the N_0 equidistant points $\overline{u}_i = -u_{\max} + 2u_{\max}\frac{i-1}{N_0-1}$, let us denote by $\overline{R}_\Lambda(u)$ the piecewise linear model of $R(u)$, spanned on the nodes

$$\left\{(\overline{u}_i, R(\overline{u}_i))\right\}_{i\in\Lambda},$$

where $\Lambda = \{1, 2, ..., N_0\}$. Let moreover define its approximation error as

$$E(\Lambda) = \int_{-u_{\max}}^{u_{\max}} \left(\overline{R}_\Lambda(u) - R(u)\right)^2 du. \tag{7.38}$$

It can easily be shown that the upper bound of the error (7.38) depends on N_0 in the following way

$$E(\Lambda) \leq \frac{8u_{\max}^3}{(N_0 - 1)^2} L^2, \tag{7.39}$$

where $L = \max(L_R, L_{\overline{R}})$ and $L_{\overline{R}}$ is the Lipschitz constant of the approximation model. In the considered case of the piecewise linear model with the nodes $\{(\overline{u}_i, R(\overline{u}_i))\}_{i \in \Lambda}$ we simply have that $L_{\overline{R}} \leq L_R$ and consequently $L = L_R$. The total approximation error of the obtained model can be evaluated as follows

$$E(\Gamma) = \int_{-u_{\max}}^{u_{\max}} \left(\overline{R}_\Gamma(u) - R(u)\right)^2 du \leq$$
$$\leq \int_{-u_{\max}}^{u_{\max}} \left(\overline{R}_\Gamma(u) - \overline{R}_\Lambda(u)\right)^2 du + \int_{-u_{\max}}^{u_{\max}} \left(\overline{R}_\Lambda(u) - R(u)\right)^2 du \leq$$
$$\leq D(\Gamma) + E(\Lambda),$$

where

$$D(\Gamma) = \frac{2u_{\max}}{N_0 - 1} \sum_{i=1}^{N_0-1} \max\left(\left(\overline{R}_\Gamma(\overline{u}_i) - R(\overline{u}_i)\right)^2, \left(\overline{R}_\Gamma(\overline{u}_{i+1}) - R(\overline{u}_{i+1})\right)^2\right) \leq$$
$$\leq \frac{4u_{\max}}{N_0 - 1} Q_l. \tag{7.40}$$

We emphasize that the initial error $E(\Lambda)$ can be simply computed, and $D(\Gamma)$ can be estimated by substituting unknown $R(\overline{u}_i)$'s in (7.40), with their nonparametric estimates $\widehat{R}_N(\overline{u}_i)$, i.e.,

$$\widehat{D}(\Gamma) = \frac{2u_{\max}}{N_0 - 1} \sum_{i=1}^{N_0-1} \max\left(\left(\overline{R}_\Gamma(\overline{u}_i) - \widehat{R}_N(\overline{u}_i)\right)^2, \left(\overline{R}_\Gamma(\overline{u}_{i+1}) - \widehat{R}_N(\overline{u}_i)\right)^2\right). \tag{7.41}$$

Remark 7.10. *The value of $\widehat{D}(\Gamma \backslash \{\widetilde{p}\})$ can be computed on the basis of $\widehat{D}(\Gamma)$ with the complexity $O(1)$.*

Efficient Polynomial Approximation

For approximation of the regression function $R(u)$, we apply the ordinary polynomial model

$$\widetilde{R}(u, \theta) = \theta_m u^{m-1} + \ldots + \theta_2 u + \theta_1 = \phi^T(\overline{u}_i)\theta, \tag{7.42}$$

where $\phi(u) = \left(u^{m-1}, ..., u, 1\right)^T \in R^{m \times 1}$, and $\theta = (\theta_m, ..., \theta_2, \theta_1)^T$. Since the model (7.42) is linear with the respect to the parameters θ, the least squares solution can be found in the standard way

$$\widehat{\theta} = \left(\Phi_{N_0}^T \Phi_{N_0}\right)^{-1} \Phi_{N_0}^T Y_{N_0},$$

where $\Phi_{N_0} = \left(\phi^T(\overline{u}_1), \phi^T(\overline{u}_2), ..., \phi^T(\overline{u}_{N_0})\right)^T \in R^{N_0 \times m}$, and

$$Y_{N_0} = \left(\widehat{R}_N(\overline{u}_1), \widehat{R}_N(\overline{u}_2), ..., \widehat{R}_N(\overline{u}_{N_0})\right)^T \in R^{N_0 \times 1}.$$

Obviously, on the finite interval $(-u_{\max}, u_{\max})$, the obtained model $\widetilde{R}(u, \widehat{\theta})$ is differentiable and fulfills the Lipschitz condition with the constant

$$L_{\widetilde{R}} = (m-1)\left|\widehat{\theta}_m\right| u_{\max}^{m-2} + (m-2)\left|\widehat{\theta}_{m-1}\right| u_{\max}^{m-3} ... + \left|\widehat{\theta}_2\right|,$$

which can be used in computation of the upper bound of the approximation error. We simulated Hammerstein system excited by the uniformly distributed random process $u_k \sim U(-1, 1)$, i.e., $u_{\max} = 1$. The static nonlinear block was followed by the IIR linear dynamics with the impulse response $\gamma_i = 0.5^{i+1}$, and the output of the whole Hammerstein system was disturbed by zero-mean colored random noise $z_k = 0.9 z_{k-1} + \varepsilon_k$, where $\varepsilon_k \sim U(-0.1, 0.1)$. Our aim was to find the smallest subset of indices $\Gamma^* \subset \Lambda$, for which the simplified piecewise linear model $\overline{R}_{\Gamma^*}(u)$ guarantees given accuracy $E_{\max} = 0.1$, i.e.,

$$\Gamma^* = \arg\min_{\Gamma \in S} \#\Gamma, \text{ where } S = \{\Gamma : E(\Gamma) \le E_{\max}\},$$

and $\#\Gamma$ is the cardinality of Γ. We applied the following simplified procedure.

Step 0. Compute $E(\Lambda)$ and set $\Gamma = \Lambda = \{1, 2, ..., N_0 = 21\}$, i.e., the initial model includes all points.

Step 1. Calculate the index $\widetilde{p} \in \Gamma$, the candidate to be removed from Γ, which causes the smallest possible increase of the approximation error, i.e.,

$$\widetilde{p} = \arg\min_{p \in \Gamma} \widehat{D}(\Gamma \backslash \{p\}).$$

Step 2. If $E(\Lambda) + \widehat{D}(\Gamma \backslash \{\widetilde{p}\}) < E_{\max}$, then set $\Gamma := \Gamma \backslash \{\widetilde{p}\}$ and go to *Step 1.*

The results for the polynomial and saturation true characteristics are presented in Fig. 7.8 and in Fig. 7.9, respectively.

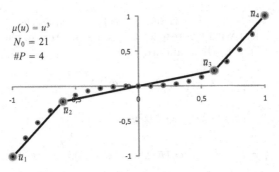

Fig. 7.8 The piecewise-linear model obtained for the polynomial characteristic $\mu(u)$

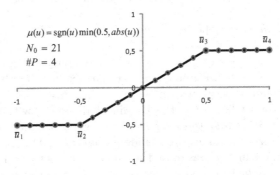

Fig. 7.9 The piecewise-linear model obtained for the saturation characteristic $\mu(u)$

7.9   Final Remarks

In the chapter we showed new algorithms, which combine the *parametric* and *nonparametric* approaches to recover the Hammerstein system nonlinearity under *quasi-parametric* (uncertain) prior knowledge. The nonparametric kernel smoothing is used first to filter the noisy learning sequence, and then, the filtered data are used to recognize the class of model. In the competition phase, the winner is the nearest, i.e., the parametric model which matches best the pattern generated by the system. One can point out the following advantages of the algorithm:

1. No need of storing N measurements in memory.
2. The random points $\{u_k\}_{k=1}^{N}$ are replaced by the freely selected deterministic ones $\{\bar{u}_i\}_{i=1}^{N_0}$, and the orthogonal plans can be executed. Moreover, the model competitions can be performed in the user-defined regions of interests, e.g., in the working points .

3. The list of model candidates can be open – the new models can join the competition at any time.
4. When the resulting parametric approximation is not satisfying, and the data-set is long enough one can apply the nonparametric correction (see [46] and [126]).

Chapter 8
Time-Varying Systems

8.1 Introduction

A large number of physical systems are nonstationary. Identification of nonstationary processes have been widely studied in the literature for linear systems. Traditional techniques for identifying linear time-varying (LTV) systems are mostly based on the recursive weighted least squares methods (see [106], [22], [83], [34], [124]). The weights are dependent on time, in the sense that the most recent measurements should be privileged, while the oldest should have the smallest influence on the estimate. If the time horizon is too long, i.e. the weights decrease too slow, we obtain the bias connected with parameter changes. On the other hand, if the horizon is short, the estimate becomes sensitive on the noise and the variance error appears. The goal is thus to design a good compromise between bias and variance, i.e., we look for a good trade-off between tracking ability and noise rejection [84], [103], [82], [104]. Some of methods proposed in the literature use the Kalman filter approach [16], or expand jump changes of the coefficients in the wavelet series [137]. As regards the identification of time-varying nonlinear block-oriented (Hammerstein and Wiener) systems, comparatively little attention has been paid in the literature. It is commonly assumed that the nonlinear regression is quasi-stationary, i.e., that the system becomes stationary as the measurement index tends to infinity (see [51], [115], [116], [117]).

In this chapter we adopt some results of adaptive modeling theory of linear systems for the nonlinear dynamic systems. In Section 8.2 we formulate the problem of nonparametric kernel regression estimation in nonstationary case. Next, in Section 8.3 we generalize the 3-stage algorithm, presented in Chapter 5, for parameter-varying NARMAX systems. In Section 8.4 we consider a special case, when the system characteristics change periodically with the known time-period T. Finally, in Section 8.5 we show the application of the model detection method, presented in Chapter 7, for time-varying systems.

G. Mzyk, *Combined Parametric-Nonparametric Identification*
of Block-Oriented Systems, Lecture Notes in Control and Information Sciences 454,
DOI: 10.1007/978-3-319-03596-3_8, © Springer International Publishing Switzerland 2014

8.2   Kernel Estimate for Time-Varying Regression

Let us first consider, for simplicity of presentation, the problem of estimation of constant value θ^*, observed in the presence of additive, zero-mean random noise z_k, with finite variance σ_z^2

$$y_k = \theta^* + z_k,$$

i.e., we recover θ^* from the observations $\{y_k\}_{k=1}^N$. In general, the measurements $\{y_k\}_{k=1}^N$ can be taken into account with different levels of significancy, i.e., we want to find the weighted least squares minimum

$$\widehat{\theta} = \arg\min_\theta \sum_{k=1}^N \alpha_k \left(y_k - \theta\right)^2, \tag{8.1}$$

where α_k's are some weights, representing the priorities of respective measurements y_k's. Since

$$\frac{\partial}{\partial \theta} \sum_{k=1}^N \alpha_k \left(y_k - \theta\right)^2 = 2 \sum_{k=1}^N \left(\theta \alpha_k - y_k \alpha_k\right),$$

the minimum in (8.1) is the solution of the equation

$$\theta \sum_{k=1}^N \alpha_k = \sum_{k=1}^N y_k \alpha_k,$$

which leads to

$$\widehat{\theta} = \frac{\sum_{k=1}^N y_k \alpha_k}{\sum_{k=1}^N \alpha_k}, \tag{8.2}$$

where $\sum_{k=1}^N \alpha_k$ can be understand as effective number of measurements. In particular, if α_k's are the same for each k, i.e., $\alpha_k = \alpha = \text{const.}$, then we obtain the standard least squares solution

$$\widehat{\theta} = \frac{\alpha \sum_{k=1}^N y_k}{N\alpha} = \frac{1}{N} \sum_{k=1}^N y_k.$$

Non-uniform α_k's are commonly applied due to two main reasons: (i) time-varying estimated value $\theta^*(N) = \theta_N^*$, and (ii) error-in variable problem in local regression estimation. This two situations are briefly commented below.

(i) If the estimated value θ_N^* is changing, the popular choice is to take

$$\alpha_k = \lambda^{N-k}, \text{ with } 0 < \lambda < 1, \tag{8.3}$$

which damps old measurements and increases the influence of recent observations on the resulting estimate. Moreover, such a choice is very convenient for designing of the recursive versions of identification algorithms, since

$$\alpha_{k+1} = \frac{1}{\lambda} \alpha_k.$$

On the other hand, since the effective number of measurements is finite, i.e. $\sum_{k=1}^{\infty} \alpha_k = \sum_{k=1}^{\infty} \lambda^{N-k} < \infty$, the variance of the estimate (8.2) does not tend to zero. It is a cost paid for good tracking abilities of the algorithm. The most popular models of θ_N^* variations are random walk, random walk with trends, jump changes, Markov chains and knowledge-based descriptions (for details see [84] and the references cited therein).

Also the restrictive (hard) selection of the fixed number n of the last measurements can be applied, i.e.,

$$\alpha_k = \begin{cases} 1, & \text{as } N - n < k \leq N \\ 0, & \text{as } k \leq N - n \end{cases}.$$

The estimate $\widehat{\theta}_n^{(N)}$, selecting n last observations from the N element data set, is then as follows

$$\widehat{\theta}_n^{(N)} = \frac{1}{n} \sum_{k=N-n+1}^{N} y_k.$$

Let us assume that the true value of estimated parameter θ^* jumps at the time instant N, from θ^* to $\theta^* + \Delta$. The moments of the measurement noise z_k remain the same. Let us also introduce the following quality index of the tracking procedure

$$Q(n) = \sum_{i=N+1}^{N+H} \left(\text{var}\widehat{\theta}_n^{(i)} + \text{bias}^2 \widehat{\theta}_n^{(i)} \right),$$

where H denotes horizon of the tracking. We simply get

$$Q(n) = \frac{H\sigma_z^2}{n} + \frac{\Delta^2}{n^2} \sum_{j=1}^{n-1} j^2 = \frac{H\sigma_z^2}{n} + \frac{\Delta^2}{n^2} \left(\frac{n(n+1)(2n+1)}{6} - n^2 \right)$$

$$\simeq \frac{H\sigma_z^2}{n} + \frac{\Delta^2}{3} n,$$

where the cumulated variance component $\frac{H\sigma_z^2}{n}$ dominates for small n, while the cumulated bias error $\frac{\Delta^2}{3} n$ increases for large n (see Fig. 8.1). The optimal value of n

$$n_{opt.} = \arg \min_n Q(n)$$

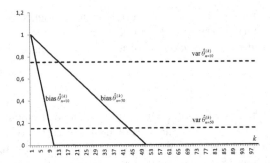

Fig. 8.1 Sample behaviour of bias and variance after parameter jump for $n = 10$ and $n = 50$

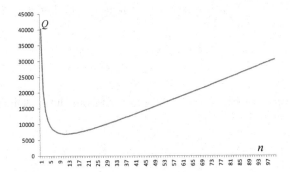

Fig. 8.2 Dependence between tracking error Q and the window length n, for $H = 100$, $\sigma_z = 20$, and $\Delta = 30$

obviously depends on the relation between horizon H, height of jump Δ, and the variance of the noise σ_z^2, i.e., $n_{opt.} = n_{opt.}(H, \Delta, \sigma_z^2)$. Nevertheless, this dependence is weak, in the sense that in typical cases (where Δ and σ_z^2 are of comparable order) n_{opt} usually lays between 10 and 20 (see Example in Fig. 8.2). Let us also emphasize that since Δ and σ_z^2 are unknown, computing $n_{opt.}$ is not possible. Also the asymptotic convergence of the estimate cannot be achieved, because of finite number of effective measurements.

(ii) In the traditional kernel regression function estimation we recover the value of time-invariant (stationary) function $\theta^* = \mu(u)$ for a given point u, using the pairs of observations $\{(u_k, y_k)\}_{k=1}^{N}$, where u_k's are random (not necessary equal to u) and

$$y_k = \mu(u_k) + z_k.$$

Assuming that $\mu()$ is a Lipschitz function, the pairs (u_k, y_k) are taken into account with different levels of significancy, and the following minimization is performed

$$\widehat{\mu}(u) = \arg\min_{\theta} \sum_{k=1}^{N} \alpha_k \left(y_k - \theta\right)^2, \tag{8.4}$$

where α_k's are some weights representing the priorities of respective measurements (u_k, y_k)'s. The weights α_k's are here dependent on the distance between kth input observation u_k, and the given estimation point u. For example we can set

$$\alpha_k = K\left(\frac{u_k - u}{h}\right), \tag{8.5}$$

where $K()$ is a kernel function, and h – a bandwidth parameter.

For a *non-stationary and nonlinear block* $\theta_N^* = \mu_N(u)$, i.e., when both problems discussed in (i) and (ii) take place, the estimated value θ^* changes in time (N is the time instant in which we want to estimate θ^*), we propose to combine (8.3) with (8.5) in the following way

$$\alpha_k = \lambda^{N-k} K\left(\frac{u_k - u}{h}\right).$$

It leads to the following, nonparametric tracking procedure

$$\widehat{\theta}_N = \widehat{\mu}_N(u) = \frac{\sum_{k=1}^{N} y_k \lambda^{N-k} K\left(\frac{u_k-u}{h}\right)}{\sum_{k=1}^{N} \lambda^{N-k} K\left(\frac{u_k-u}{h}\right)}. \tag{8.6}$$

The generalized kernel estimate (8.6) with the forgetting factor λ can be applied in the regression estimation in Hammerstein systems (see (2.12)), inverse regression estimation in Wiener systems (see (3.2)), and in the censored sample mean approach to Wiener-Hammerstein systems (see (4.5)). It can also support generation of instrumental variables for the IIR Hammerstein/NARMAX systems (see (2.31) and (5.55)), and compress the data in the 3-step procedure of model recognition (see page 158).

Below, we present the results of simple experiment, in which the time-varying static characteristic $\mu_N(u) = u^2 + c_N$, with jumping offset $c_N = 1_{N-30}$ was recovered in the point $u = 0$, under random input $u_k \sim U[-1,1]$ and in the presence of random output noise $z_k \sim U[-0.1, 0.1]$. As can be seen in Fig. 8.3, for small value of λ (e.g., $\lambda = 0.6$) we observe rapid reaction with huge variance of the results. For λ close to 1 the variance is reduced, but it is done at the expense of inertia.

8.3   Weighted Least Squares for NARMAX Systems

In this point we analyze applicability of the recursive weighted least squares and instrumental variables algorithms and the modified kernel regression method (8.6) in the 3-Stage identification of additive NARMAX system (see Section 5.7) and Hammerstein system, as its particular case (see

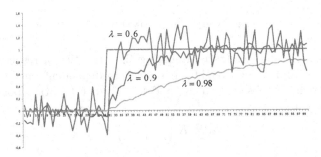

Fig. 8.3 Tracking results for various values of the forgetting factor λ

Appendix A.3). Below, we present the stages of the procedure for time-varying NARMAX systems.

Stage 1 (nonparametric). Generate empirical matrix of instruments $\widehat{\Psi}^*_{N,M} = (\widehat{\psi}^*_{1,M}, \widehat{\psi}^*_{2,M}, ... \widehat{\psi}^*_{N,M})^T$, where

$$\widehat{\psi}^*_{k,M} = (f_1(u_k), ..., f_m(u_k), ..., f_1(u_{k-n}), ... \tag{8.7}$$
$$..., f_m(u_{k-n}), \widehat{R}_{1,M}(u_{k-1}), ..., \widehat{R}_{q,M}(u_{k-1}), ..., \widehat{R}_{1,M}(u_{k-p}), ..., \widehat{R}_{q,M}(u_{k-p}))^T,$$

and

$$\widehat{R}_{l,M}(u_j) = \frac{\sum_{i=1}^{j}\left(g_l(y_i)\lambda^{j-i}K(\frac{u-u_i}{h(M)})\right)}{\sum_{i=1}^{j}\lambda^{j-i}K(\frac{u-u_i}{h(M)})},$$

with $0 < \lambda < 1$.

Stage 2 (parametric). Estimate the aggregated parameter vector (5.8) using the recursive least squares or instrumental variables method

$$\widehat{\theta}^{(LS)}_k = \widehat{\theta}^{(LS)}_{k-1} + P^{(LS)}_k \phi_k(y_k - \phi_k^T\widehat{\theta}^{(LS)}_{k-1}), \tag{8.8}$$
$$\widehat{\theta}^{(IV)}_k = \widehat{\theta}^{(IV)}_{k-1} + P^{(IV)}_k \psi_k(y_k - \phi_k^T\widehat{\theta}^{(IV)}_{k-1}),$$

with

$$P^{(LS)}_k = P^{(LS)}_{k-1} - \frac{1}{1 + \phi_k^T P^{(LS)}_{k-1}\phi_k}P^{(LS)}_{k-1}\phi_k\phi_k^T P^{(LS)}_{k-1}, \tag{8.9}$$

$$P^{(IV)}_k = P^{(IV)}_{k-1} - \frac{1}{1 + \phi_k^T P^{(IV)}_{k-1}\psi_k}P^{(IV)}_{k-1}\psi_k\phi_k^T P^{(IV)}_{k-1},$$

or minimize the weighted criterion (see [84])

$$\sum_{k=1}^{N}\alpha_k\left(y_k - \phi_k^T\theta\right)^2 \to \min_{\theta},$$

in the following (recursive) way

$$\widehat{\theta}_k^{(LS)} = \widehat{\theta}_{k-1}^{(LS)} + L_k^{(LS)}(y_k - \phi_k^T \widehat{\theta}_{k-1}^{(LS)}),$$ $$\quad (8.10)$$
$$\widehat{\theta}_k^{(IV)} = \widehat{\theta}_{k-1}^{(IV)} + L_k^{(IV)}(y_k - \phi_k^T \widehat{\theta}_{k-1}^{(IV)}),$$

where

$$L_k^{(LS)} = \frac{P_{k-1}^{(LS)} \phi_k}{\alpha_k + \phi_k^T P_{k-1}^{(LS)} \phi_k}, \qquad L_k^{(IV)} = \frac{P_{k-1}^{(IV)} \phi_k}{\alpha_k + \psi_k^T P_{k-1}^{(IV)} \phi_k},$$

and

$$P_k^{(LS)} = \frac{1}{\alpha_k} \left[P_{k-1}^{(LS)} - \frac{P_{k-1}^{(LS)} \phi_k \psi_k^T P_{k-1}^{(LS)}}{\alpha_k + \psi_k^T P_{k-1}^{(LS)} \phi_k} \right],$$

$$P_k^{(IV)} = \frac{1}{\alpha_k} \left[P_{k-1}^{(IV)} - \frac{P_{k-1}^{(IV)} \phi_k \psi_k^T P_{k-1}^{(IV)}}{\alpha_k + \psi_k^T P_{k-1}^{(LS)} \phi_k} \right].$$

Stage 3 (decomposition). Similarly as for time-invariant system compute the SVD (singular value decomposition) of the matrices $\widehat{\Theta}_{\lambda d}^{(IV)}$ and $\widehat{\Theta}_{\gamma c}^{(IV)}$, i.e., $\widehat{\Theta}_{\gamma c}^{(IV)} = \sum_{i=1}^{\min(n,m)} \sigma_i \widehat{\mu}_i \widehat{\nu}_i^T$, $\widehat{\Theta}_{\lambda d}^{(IV)} = \sum_{i=1}^{\min(p,q)} \delta_i \widehat{\xi}_i \widehat{\zeta}_i^T$ to obtain the estimates of changing parameters (see (5.65) on page 132).

8.4 Nonparametric Identification of Periodically Varying Systems

Let us consider the continuous-time Hammerstein system with periodically varying nonlinear static characteristic with a priori known period T. The goal is to estimate periodic regression

$$R(u,t) = R(u, t + T)$$

for each $t \in [0, T]$, from the randomly sampled measurements $\{(t_k, u_k, y_k)\}$, where t_k is the time instance of the collected pair (u_k, y_k). We propose the following kernel method

$$\widehat{R}_N(u,t) = \frac{\sum_{k=1}^N y_k K\left(\frac{1}{h_N} \left\| \begin{bmatrix} u_k \\ t_k^* \end{bmatrix} - \begin{bmatrix} u \\ t \end{bmatrix} \right\| \right)}{\sum_{k=1}^N K\left(\frac{1}{h_N} \left\| \begin{bmatrix} u_k \\ t_k^* \end{bmatrix} - \begin{bmatrix} u \\ t \end{bmatrix} \right\| \right)},$$

in which

$$t_k^* = t_k - n_k T, \text{ and } n_k = \left\lfloor \frac{t_k}{T} \right\rfloor.$$

Let $f(u,t)$ be joint probability density function of the input u and the time t. The following theorem holds.

Theorem 8.1. *If $h_N \to 0$ and $N h_N^2 \to \infty$, as $N \to \infty$, and moreover both $R()$ and $f()$ are continuous in the point (u,t), then*

$$\widehat{R}_N(u,t) \to R(u,t) \tag{8.11}$$

provided that $f(u,t) > 0$.

Proof. The conclusion (8.11) follows directly from Theorem 1 in [45]. ∎

8.5 Detection of Structure Changes

The modified kernel estimate for time-varying regression (see (8.6)) can also be applied in the first step of the procedure of model selection, described in detail in Chapter 7. Obviously, when the true system characteristic changes in time, then, the index of optimal approximating class is also function of time. The only needed modification is to replace the standard kernel estimate (7.18) (see *Step 1* on page 7.5), with the weighted approach, given by (8.6). Steps 2 and 3 of the procedure remain unchanged. Such a strategy seems to be an interesting proposition in switched system control and fault detection. Nevertheless, since the effective number of measurements in (8.6) is finite for $|\lambda| < 1$ (even asymptotically) the results given by Theorem 7.1 and Remark 7.8 cannot be guaranteed, i.e., we have non-zero risk of bad classification. Thus, let us consider a special case of the class of time-varying nonlinearities given by assumption below.

Assumption 8.1. *The nonlinear characteristic $\mu()$ switches between $\mu_1()$ and $\mu_2()$ at time $t = t_0$, i.e.,*

$$\mu(u,t) = \begin{cases} \mu_1(u), & \text{as } t < t_0 \\ \mu_2(u), & \text{as } t \geq t_0 \end{cases} \tag{8.12}$$

and, moreover, both $\mu_1()$ and $\mu_2()$ fulfill Assumption 7.2.

The following theorem holds.

Theorem 8.2. *If the time-varying static characteristic $\mu()$ of Hammerstein system fulfills Assumption 8.1 then the 3-step algorithm (see (7.27), (7.28), (7.29), and (7.30) on page 159) tends to the best model of $\mu_2(u)$ with probability 1, as $N \to \infty$.*

Proof. The proof is a simple consequence of Theorem 7.1 and the fact that we can put $\lambda = 1$ in (8.6), thanks to specific (quasi-stationary) behaviour of $\mu(u, t)$ given by (8.12). The convergence

$$\widehat{R}_N(\overline{u}_i) \to R_2(\overline{u}_i), \qquad i = 1, 2, ..., N_0,$$

where $R_2(\overline{u}_i) = E\{y_k | u_k = u, t \geq t_0\}$ is the regression in Hammerstein system after jump, follows from Lemma 1 in [51]. ∎

The Theorem 8.2 is illustrated below, by the simple experiment. The Hammerstein system with excitations and the dynamic part as in Section 7.8.1 and switched nonlinear characteristic (see (8.12))

$$\mu_1(u) = c_{1,1}^* \left(e^{c_{1,2}^* u} - 1 \right), \quad \text{and} \quad \mu_2(u) = c_{2,1}^* u + c_{2,2}^*,$$

was simulated. We assumed sampling time $T = 1$ and $t_0 = 100$. In Fig. 8.4 we present the index of the true class (solid line) and the decisions obtained in (7.30) (dots) for $\lambda = 1$ and $\lambda = 0.9$ applied in (8.6), respectively. According to intuition, for $\lambda = 1$ and N large, we have correct classification with the probability close to one. Nevertheless, reaction on switch is relatively slow. The delay is connected with the fact that the number of new measurements (after switch) is comparable with the number of old data not before $N \geq 200$. On the other hand, for $\lambda = 0.9$ reaction is rapid, since the weights of new measurements dominate already for $N \geq 107$. Unfortunately, the measurement noise causes high risk of bad classification, even asymptotically.

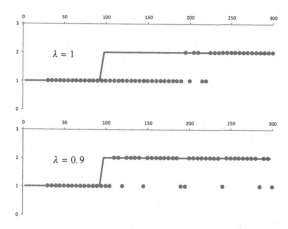

Fig. 8.4 Structure detection of switching system for various forgetting factors λ

8.6   Final Remarks

Thanks to modular structure of the combined parametric-nonparametric algorithms, where the individual stages realize standard procedures, the ideas of identification of simple time-varying objects can be adopted for complex systems in a very convenient and natural way. In particular, the system components are identified by the standard weighted least squares or kernel regression procedures. Obviously, since the system changes in time, the consistency of estimates cannot be expected, with the exception of periodically varying systems and quasi-stationary systems, discussed in Sections 8.4 and 8.5, respectively.

Chapter 9
Simulation Studies

9.1 IFAC Wiener-Hammerstein Benchmark

In this chapter we demonstrate cooperation between recent nonparametric and traditional nonparametric methods in system identification. The goal is not to show results of competition of both groups, but to demonstrate promising possibilities of cooperation.

Since the parametric and nonparametric methods work under extremely different assumptions, any comparisons of this two kinds of approaches are not possible and have no sense. Assumption of parametric representation of the system (with finite and known number of unknown parameters) guarantees fast convergence of the model to the best approximation in the considered class. Nevertheless, if the assumed class is not correct, the estimate cannot overcome systematic approximation error, even for huge number of measurements. The philosophy of nonparametric algorithms is different, in the sense that they 'trust' only the measurements. Obviously, this fact can exclude the approach for short data sequences and long-memory dynamics. However, asymptotically, nonparametric estimates reach the true characteristic of the system. Consequently, in some situations parametric approach can give better results, while in the other, the results can be improved by nonparametric correction.

As it will be shown in the experiments below, computing nonparametric estimates in the preliminary step of the identification procedure can support selection of the parametric class of models. In particular, the censored sample mean algorithm (see (3.14) and (4.5)) can be applied *without any prior knowledge of the system* on the raw input-output data $\{(u_k, y_k)\}$ for nonparametric recovering of the static nonlinear characteristic. The knowledge of the shape of nonlinear characteristic can help to choose proper basis functions in its parametric representation (e.g. piecewise linear, or polynomial).

Selected methods proposed in the book was tested on the real data obtained from electronic circuit, including the nonlinear element (diode), and surrounded by two linear dynamic filters (L-N-L sandwich). For detail

G. Mzyk, *Combined Parametric-Nonparametric Identification*
of Block-Oriented Systems, Lecture Notes in Control and Information Sciences 454,
DOI: 10.1007/978-3-319-03596-3_9, © Springer International Publishing Switzerland 2014

description of the used IFAC Wiener-Hammerstein benchmark we refer the reader to [122]. Below, we discuss some aspects which can prefer or depreciate some methods in this particular problem.

- As it was mentioned before, the most important advantage of the non-parametric stages proposed in the book is that they can be applied with the lack or poor prior knowledge of the characteristic. This fact is important especially for strong, irregular nonlinearities, while in the considered benchmark, the nonlinear characteristic is nearly linear over some regions, and traditional methods can give acceptable results in practice.
- Nonparametric instrumental variables, proposed in Section 2.2.3, can significantly reduce the bias for random, strongly correlated and heavy disturbances, while in the considered benchmark the noise is relatively small. Consequently, this advantage cannot be demonstrated on the benchmark data.
- The strong point of the nonparametric, censored sample mean approach, proposed and analyzed in Section 3.2.3, is that the true nonlinear characteristic of Wiener-Hammerstein system can be recovered for any distribution of excitation. In the IFAC benchmark, the Gaussian input is generated, which does not exclude applicability of the traditional methods. Obviously, weaker assumptions imposed on nonparametric algorithms proposed in the book are also fulfilled, but since this methods are more general and universal, they can give worse results for specific systems.

9.2   Combined Methods in Action

We start our identification procedure from the standard input-output cross-correlation analysis (see (1.23) and (3.5)). The results are shown in Fig. 9.1. Next, applying standard numerical procedures for the obtained sequence $C_{y,u}(\tau)$ we conclude that it can be upper-bounded as follows

$$|C_{y,u}(\tau)| < c\lambda^\tau, \text{ with } \lambda > 0.892, \text{ and some unknown } 0 < c < \infty.$$

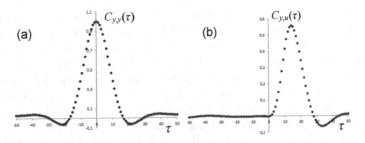

Fig. 9.1 Output autocorrelation function (a) and input-output cross-correlation function (b)

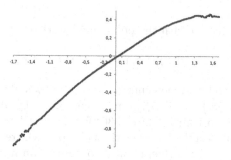

Fig. 9.2 The shape of the static nonlinear characteristic obtained by nonparametric censored sample mean method

In the second step, the nonparametric censored sample mean algorithm (4.5) is run with the tuning parameter $\lambda = 0.9$ for recovering of the static nonlinear characteristic of the system. The result shown in Fig. 9.2 is built on the basis of the measurements only and gives general view on the shape of identified function. Resulting points can be used in the model recognition method presented in Chapter 7. Below, we present the list of compared models and shortly describe each of them.

A. Linear FIR(40) moving average model

$$\overline{y}_k = \sum_{i=0}^{40} \widehat{\gamma}_i u_{k-i}, \tag{9.1}$$

with the coefficients $\{\widehat{\gamma}_i\}_{i=0}^{40}$ obtained by the standard nonparametric correlation-based algorithm (1.23).

B. Hammerstein model with two-segment characteristic

$$\overline{\mu}(u) = \begin{cases} c_1 u, & \text{as } u \geq u_p \\ c_2 u, & \text{as } u < u_p \end{cases} \tag{9.2}$$

with the border point $u_p = 0$, followed by the FIR(40) linear filter. The parameters $\{c_i\}_{i=1}^{2}$ and $\{\widehat{\gamma}_i\}_{i=0}^{40}$ computed by the 3-stage algorithm (5.19).

C. Model B. with $u_p = 0.5$.

D. Model B. with $u_p = 1$.

E. Hammerstein model with two-segment piecewise linear characteristic

$$\overline{\mu}(u) = \begin{cases} c_1 u + c_2, & \text{as } u \geq u_p \\ c_3 u + c_4, & \text{as } u < u_p \end{cases} \tag{9.3}$$

with the node in $u_p = 0$, followed by the FIR(40) linear filter. The parameters $\{c_i\}_{i=1}^{4}$ and $\{\widehat{\gamma}_i\}_{i=0}^{40}$ computed by the 3-stage algorithm (5.19).

F. Model E. with $u_p = 0.3$.

G. Model E. with $u_p = 0.5$.

H. Model E. with $u_p = 1$.

I. L-N-L sandwich model with the polynomial static characteristic

$$v = \overline{\mu}(x) = c_1 x^3 + c_2 x^2 + c_3 x + c_4$$

surrounded by two linear elements $x_k = ax_{k-1} + u_k$ and $\overline{y}_k = \sum_{i=0}^{40} \widehat{\gamma}_i v_{k-i}$. The parameters $\{c_i\}_{i=1}^{4}$, $\{\widehat{\gamma}_i\}_{i=0}^{40}$ and a computed by alternate updating with the use of algorithms (3.29)-(3.30) for the first linear subsystem and the instrumental variables based method (5.19) for the remaining Hammerstein part. The two-level (hierarchical) optimization method was applied ([32])

$$\min_{a,c,\gamma} \left\| Y_N - \overline{Y}(a,c,\gamma) \right\| = \min_a \min_{c,\gamma} \left\| Y_N - \overline{Y}(a,c,\gamma) \right\|.$$

J. Model I. with the piecewise linear nonlinearity (9.3) and $u_p = 0.54$ indicated by the algorithm described in Section 7.8.3.

K. NARMAX model (5.1) consisting of two Hammerstein branches with the AR(1) linear subsystems, and the nonlinearities as in (9.3). This model uses true values of the previous system output measurements, which is *not allowed* in the benchmark rules described in [122].

9.3 Results of Experiments

According to recommendations given in [122], the following errors were computed for each algorithm:

- the mean absolute error

$$ERR = \frac{1}{87000} \sum_{k=101001}^{188000} |y_k - \widehat{y}_k|,$$

- the root mean square error

$$RMS = \sqrt{\frac{1}{87000} \sum_{k=101001}^{188000} (y_k - \widehat{y}_k)^2}.$$

The results are collected in Table 10.1 and presented in Fig. 9.3. Figures 9.4 and 9.5 shows the fragment of system output, and illustrates the refinement resulting from application of nonlinear sandwich model, with properly selected basis functions. Figure 9.6 illustrates the obtained errors of models I. and J. for various values of a.

Table 9.1 Comparison of the mean absolute errors and the root mean errors of various models and algorithms

Model / algorithm	Mean absolute error ERR $[\cdot 10^{-4}]$	Root mean error RME $[\cdot 10^{-4}]$
A	403	565
B	240	331
C	222	310
D	199	281
E	207	284
F	188	259
G	181	252
H	190	267
I	129	175
J	118	168
K	63	75

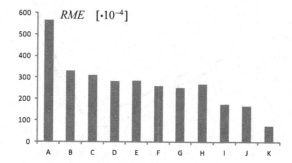

Fig. 9.3 Root mean error of various models and algorithms

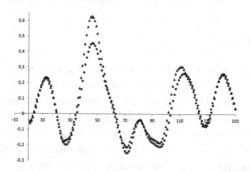

Fig. 9.4 System output (balls) and the linear FIR(40) model output (triangles)

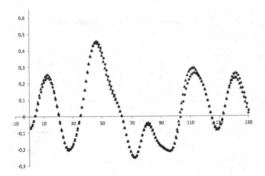

Fig. 9.5 System output (balls) and Wiener-Hammerstein piecewise-linear model output (triangles)

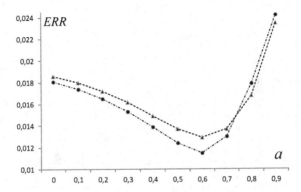

Fig. 9.6 Sandwich model error for various values of parameter of the AR(1) input filter with polynomial (triangles) and piecewise-linear (balls) nonlinearity

9.4 Conclusions

The nonparametric censored algorithm (4.5) is a very useful tool, which allows to recover the shape of the static nonlinear characteristic without any prior knowledge about the system. The result can support model recognition. Nevertheless, the estimate has some limitations in practical use. Firstly, there is no formal rules of selecting the parameter λ, needed to be set in (4.5), and further studies are needed to elaborate automatic tuning. Secondly, the variance of the nonlinearity estimate increases significantly in the regions of small value of the input probability density function (see e.g. the results in Fig. 9.2 for $|u_k| > 1.5$). This fact is influential especially for moderate number of observations.

In general, incorporation of nonlinear component into the linear model A allows to improve the accuracy, with the extent dependent on the used structure and the type of nonlinearity. The algorithms using Hammerstein models (i.e. B÷H) reduce the error, but the correction is not significant. Since the Hammerstein model includes only one linear block, it cannot provide good approximation of the input linear dynamic filter of the real system. Similar situation takes place for Wiener models.

Better results are obtained for sandwich model, which represents the true structure of interconnections in the real system. Moreover, application of the piecewise linear function J instead of the polynomial model I, including the same number of parameters, leads to additional 10% reduction of the error.

As regards the NARMAX model K, its perfect approximation is a consequence of the additional knowledge, and thus it, should not be compared with A÷K.

The experiment shows that parametric and nonparametric algorithms can be selected and combined in a very flexible and convenient way. Resulting strategies are more general and universal than 1-stage procedures, i.e. the scope of its applicability is wider. Simultaneously, even in standard parametric problems they can give comparable results.

Chapter 10
Summary

All methods elaborated in the book combine the nonparametric and parametric tools. Such a strategy allows to solve various kinds of specific obstacles, which are difficult to be overcome in purely parametric or purely nonparametric approach. In particular, the global identification problem can be decomposed on simpler local problems, the measurement sequence can be pre-filtered in the nonparametric stage, or the rough parametric model can be refined by the nonparametric correction when the number of measurements is large enough. The schemes proposed in the book can be used elastically and have a lot of degrees of freedom. In most of them we can obtain traditional parametric or nonparametric procedures by simple avoiding of the selected steps of combined algorithms. In this sense, the proposed ideas can be treated as generalizations of classical approaches to system identification. Below, to summarize the contribution, in Section 10.1 we itemize the most significant problems solved in the book, and next, in Section 10.2 we discuss selected problems remained for further studies.

10.1   Index of Considered Problems

Partial Prior Knowledge. It is possible to identify parameters of one subsystem with no prior knowledge of the remaining components of complex system (Sections 2.1, 2.2, and 3.3).

Uncertain Prior Knowledge. Thanks to the nonparametric decomposition, the algorithms are robust on the false assumptions imposed on complementary blocks (Sections 2.7, and 7.5).

Correlated Input Process. Traditional methods of deconvolution can support generation of instrumental variables (Sections 2.4, and 2.5).

Correlated Disturbances. The scope of usability of instrumental variables technique was significantly extended for a broad class of nonlinear systems (Section 2.4.4).

G. Mzyk, *Combined Parametric-Nonparametric Identification*   189
of Block-Oriented Systems, Lecture Notes in Control and Information Sciences 454,
DOI: 10.1007/978-3-319-03596-3_10, © Springer International Publishing Switzerland 2014

Generation of Instrumental Variables. The nonparametric procedures of instruments generation were proposed (Sections 2.2.3, 2.4.6, 2.5.4, 2.7.2, and 5.6).

Infinite Impulse Response. Only asymptotic stability of the linear dynamic blocks is assumed and, ARMA, AR and ARX objects are admitted as special cases (Sections 2.2.3, and 3.2.3).

Characteristics Nonlinear in the Parameters. It is possible to identify static characteristics of block-oriented systems, which are not linear in the parameters (Section 2.3).

Non-Gaussian Input. The standard Gaussianity assumption concerning the random excitation of Wiener system was relaxed to any distribution (Sections 3.2, and 3.2.3).

Non-invertible Characteristics. The class of invertible static characteristics commonly assumed for nonparametric Wiener systems was generalized for the Lipschitz functions (Section 3.2.3).

Small Number of Data. The parametric estimate can be incorporated into the kernel nonparametric algorithm (semiparametric approach). For small number of observations, the identification error can be reduced in comparison with purely nonparametric algorithm.(Section 2.6, 2.7).

Unknown Class of Model. Nonparametric methods can play the role of generator of compressed pattern of the system used for fast model recognition or model order selection (Sections 7.3, 7.4, and 7.5).

Unknown Structure. Universal algorithms which recover nonlinear characteristic, independently of the system structure, were proposed (Section 4.4).

Inaccessibility of Interconnecting Signals. Nonparametric estimation of hidden signals allows to decompose the global identification problem and, moreover, helps to generate instrumental variables (Section 2.1, 2.2, 7.2).

Time-Varying Characteristics. Typical algorithms of parameter tracking (e.g. recursive weighted least squares) can be simply adopted in the proposed multi-stage procedures (Sections 8.2, 8.3, 8.4, and 8.5).

Block-Oriented Structures with Feedback. The nonlinear dynamic additive NARMAX systems can be identified even under correlated excitations (Sections 5.6, and 5.6).

Numerical Problems. It is possible to improve, to some extent, numerical properties of the algorithms, by e.g. proper selection of estimation points (Sections 2.3.3, 3.3, 4.5, 5.2, 6.4, 7.5).

10.2   A Critical Look at the Combined Approach

Now, we are in the position to indicate some drawbacks and problems, which are not solved in the book, and are left for future research.

A critical meaning has the fact that all nonparametric methods are very sensitive on tuning parameters (scale, bandwidth), especially for moderate number of measurements (see e.g. [50], [92]). In consequence, we can obtain different results depending on the applied settings in the first stage (e.g., type of kernel or basis functions). Then, since the proposed procedures have two or three steps, the errors resulting from bad configuration on step one propagate on the next steps, and, in general can be amplified.

In the whole book we assume that the structure of interconnections is known a priori. In practice, when the arbitrary pairs (u_k, y_k) are processed we always obtain approximation error resulting from the selected model. We refer the reader to [1], [11] and [54], where the algorithms of structure detection are presented.

The issue of confidence bands and distribution of parameter estimates for finite number of measurements is not considered in the book. The analysis of uncertainty is very difficult under poor prior knowledge about the system, assumed in the book.

The rate of convergence is very sensitive on the dimension of the input signal. Recent algorithms (see [148]) allows to speed up the convergence for some special cases of input sequences.

We emphasize that, till now, the necessary and sufficient identifiability conditions for Wiener system have not been formulated and only special cases are considered in the literature.

Since the presented methods do not admit the input noises, the accurate system excitation $\{u_k\}$ must be observed. In general, the error in variable problem was not considered in the book.

Appendix A
Proofs of Theorems and Lemmas

A.1 Recursive Least Squares

The standard off-line version of the least squares estimate

$$a_N = (X_N^T X_N)^{-1} \cdot X_N^T Y_N \qquad (A.1)$$

can be transformed to the recursive version

$$a_N = f(a_{N-1}, x_N, y_N) = a_{N-1} + \delta(a_{N-1}, x_N, y_N),$$

where the correction $\delta(a_{N-1}, x_N, y_N)$ is computed online, with no need of storing measurements in memory. We can simply decompose the second term in (A.1) as

$$X_N^T Y_N = \sum_{k=1}^{N} x_k y_k = \sum_{k=1}^{N-1} x_k y_k + x_N y_N = X_{N-1}^T Y_{N-1} + x_N y_N.$$

Also the matrix $X_N^T X_N$ can be decomposed in a similar way

$$X_N^T X_N = \sum_{k=1}^{N} x_k x_k^T = \sum_{k=1}^{N-1} x_k x_k^T + x_N x_N^T = X_{N-1}^T X_{N-1} + x_N x_N^T.$$

but we need the relation between $(X_N^T X_N)^{-1}$ and $(X_{N-1}^T X_{N-1})^{-1}$. Let us denote

$$P_N = (X_N^T X_N)^{-1},$$

and observe that

$$a_N = P_N X_N^T Y_N,$$
$$a_{N-1} = P_{N-1} X_{N-1}^T Y_{N-1},$$

where
$$P_{N-1}^{-1} = X_{N-1}^T X_{N-1} \text{ and } P_N = \left(P_{N-1}^{-1} + x_N x_N^T\right)^{-1}.$$

The lemma below allows to obtain relation between P_N and P_{N-1}.

Lemma A.1. *[128] Let A be square matrix and u be a vector of the same dimension. If A^{-1} and $\left(A + uu^T\right)^{-1}$ exist, then*

$$\left(A + uu^T\right)^{-1} = A^{-1} - \frac{1}{1 + u^T A^{-1} u} A^{-1} uu^T A^{-1}.$$

Using Lemma A.1, for $A = P_{N-1}^{-1}$ and $u = x_N$ we get

$$P_N = P_{N-1} - \frac{1}{1 + x_N^T P_{N-1} x_N} P_{N-1} x_N x_N^T P_{N-1} =$$
$$= P_{N-1} - \varkappa_N P_{N-1} x_N x_N^T P_{N-1},$$

where
$$\varkappa_N = \frac{1}{1 + x_N^T P_{N-1} x_N},$$

which leads to

$$a_N = \left(P_{N-1} - \varkappa_N P_{N-1} x_N x_N^T P_{N-1}\right)\left(X_{N-1}^T Y_{N-1} + x_N y_N\right) =$$
$$= P_{N-1} X_{N-1}^T Y_{N-1} + P_{N-1} x_N y_N -$$
$$- [\varkappa_N P_{N-1} x_N](x_N^T P_{N-1} X_{N-1}^T Y_{N-1} + x_N^T P_{N-1} x_N y_N) =$$
$$= a_{N-1} + [\varkappa_N P_{N-1} x_N]\varrho_N$$

where
$$\varrho_N = \frac{1}{\varkappa_N} y_N - x_N^T a_{N-1} - x_N^T P_{N-1} x_N y_N,$$

and since
$$\frac{1}{\varkappa_N} = 1 + x_N^T P_{N-1} x_N$$

we obtain that

$$\varrho_N = y_N + x_N^T P_{N-1} x_N y_N - x_N^T a_{N-1} - x_N^T P_{N-1} x_N y_N = y_N - x_N^T a_{N-1}.$$

Consequently,

$$a_N = a_{N-1} + \varkappa_N P_{N-1} x_N (y_N - x_N^T a_{N-1}) \tag{A.2}$$
$$P_N = P_{N-1} - \varkappa_N P_{N-1} x_N x_N^T P_{N-1}$$

and

$$P_N x_N = P_{N-1} x_N - \varkappa_N P_{N-1} x_N x_N^T P_{N-1} x_N =$$
$$= \varkappa_N P_{N-1} x_N \{ \frac{1}{\varkappa_N} - x_N^T P_{N-1} x_N \},$$

where $\{ \frac{1}{\varkappa_N} - x_N^T P_{N-1} x_N \} = 1$, i.e.,

$$P_N x_N = \varkappa_N P_{N-1} x_N. \tag{A.3}$$

By inserting (A.3) to (A.2) we obtain finally

$$a_N = a_{N-1} + P_N x_N (y_N - x_N^T a_{N-1}),$$
$$P_N = P_{N-1} - \frac{P_{N-1} x_N x_N^T P_{N-1}}{1 + x_N^T P_{N-1} x_N}.$$

A.2   Covariance Matrix of the LS Estimate

Since

$$\widehat{a}_N - a^* = L_N Z_N$$

the covariance matrix has the form

$$cov(\widehat{a}_N) = E \left\{ (\widehat{a}_N - a^*)(\widehat{a}_N - a^*)^T \right\} =$$
$$= E \left\{ L_N Z_N Z_N^T L_N^T \right\} = L_N \left(E Z_N Z_N^T \right) L_N^T,$$

where

$$E Z_N Z_N^T = \begin{bmatrix} E z_1^2 & E z_1 z_2 & .. & E z_1 z_N \\ E z_2 z_1 & E z_2^2 & .. & : \\ : & : & : & : \\ E z_N z_1 & .. & & .. & E z_N^2 \end{bmatrix} = \sigma_z^2 I.$$

Thus,

$$cov(\widehat{a}_N) = \sigma_z^2 L_N L_N^T,$$

where

$$L_N L_N^T = (X_N^T X_N)^{-1} X_N^T X_N (X_N^T X_N)^{-1} = (X_N^T X_N)^{-1},$$

and consequently

$$cov(\widehat{a}_N) = \sigma_z^2 (X_N^T X_N)^{-1}.$$

Let us define the class of linear unbiased estimates (LUE) as

$$\{ \alpha_N : \alpha_N = M_N Y_N, \quad E\alpha_N = a^*, \quad M_N - \text{deterministic} \}.$$

Since

$$E\alpha_N = E M_N Y_N = E M_N X_N a^* + E M_N Z_N$$

we conclude that

$$M_N X_N = I$$

and the covariance matrix of the estimate α_N can be expressed as

$$cov(\alpha_N) = E\left\{M_N Z_N Z_N^T M_N^T\right\} = M_N \left(EZ_N Z_N^T\right) M_N^T = \sigma_z^2 M_N M_N^T.$$

It will be shown that

$$cov(\widehat{a}_N) \le cov(\alpha_N)$$

where the symbol '$\le$' is explained in Definition A.1.

Since

$$(M_N - L_N)(M_N - L_N)^T = (M_N - L_N)(M_N^T - L_N^T) =$$
$$= M_N M_N^T - M_N L_N^T - L_N M_N^T + L_N L_N^T,$$

observing that

$$M_N L_N^T = M_N X_N \left(X_N^T X_N\right)^{-1} = I \left(X_N^T X_N\right)^{-1} = \left(X_N^T X_N\right)^{-1} = L_N L_N^T,$$
$$L_N M_N^T = (M_N L_N^T)^T = L_N L_N^T,$$

we obtain

$$(M_N - L_N)(M_N - L_N)^T = M_N M_N^T - L_N L_N^T,$$

and from Lemma A.2 we have that

$$M_N M_N^T - L_N L_N^T \geqslant 0 \qquad / \cdot \sigma_z^2$$
$$cov(\alpha_N) - cov(\widehat{a}_N) \geqslant 0.$$

Definition A.1. *Matrix $A \in R^{n,n}$ is positive semidefinited (i.e. $A \geq 0$), if for each vector $w = R^n$ it holds that*

$$w^T A w \geq 0.$$

Lemma A.2. *If A is of full rank then $AA^T \geq 0$.*

A.3   Hammerstein System as a Special Case of NARMAX System

Lemma A.3. *The additive NARMAX system with the linear function $\eta(y_k)$, i.e., of the form $\eta(y_k) = dy_k$, is equivalent to the Hammerstein system.*

Proof. The NARMAX system description

$$y_k = \sum_{j=1}^{p} a_j \eta(y_{k-j}) + \sum_{i=0}^{n} b_i \mu(u_{k-i}) + v_k,$$

for $\eta(y_k) = dy_k$ and the 'input'

$$x_k \triangleq \sum_{i=0}^{n} b_i \mu(u_{k-i}) + v_k, \tag{A.4}$$

resembles the difference equation of AR linear model

$$y_k = \sum_{j=1}^{p} a_j dy_{k-j} + x_k,$$

which can be presented equivalently as ([57])

$$y_k = \sum_{l=0}^{\infty} r_l x_{k-l}. \tag{A.5}$$

Inserting (A.4) to (A.5) we obtain that

$$y_k = \sum_{l=0}^{\infty} r_l \left(\sum_{i=0}^{n} b_i \mu(u_{k-i-l}) + v_{k-l} \right),$$

and further

$$y_k = \sum_{q=0}^{\infty} \gamma_q \mu(u_{k-q}) + z_k, \tag{A.6}$$

where $z_k = \sum_{l=0}^{\infty} r_l v_{k-l}$, $\gamma_q = \sum_{l=0}^{\infty} \sum_{i=0}^{n} r_l b_i \delta(l+i-q)$, and $\delta()$ is a discrete impulse. Equation (A.6) represents Hammerstein system with infinite impulse response. ∎

A.4 Proof of Theorem 2.1

Proof. Denote $A_{N_0} = \left(\Phi_{N_0}^T \Phi_{N_0} \right)^{-1} \Phi_{N_0}^T$. Since by (2.2) in Assumption 2.1, for each vector norm $\|\cdot\|$ and the induced matrix norm it holds that $\|A_{N_0}\| \leq C$, some $C > 0$, thus (see (2.7), (2.13))

$$\|\widehat{c}_{N_0,M} - c\| \leq C \left\| \widehat{W}_{N_0,M} - W_{N_0} \right\|.$$

By equivalence of vector norms we get further

$$\left\| \widehat{W}_{N_0,M} - W_{N_0} \right\| \leq \alpha \left\| \widehat{W}_{N_0,M} - W_{N_0} \right\|_1, \text{ some } \alpha > 0,$$

where $\|x\|_1 \triangleq \sum_{i=1}^{\dim x} |x[i]|$. Consequently (cf. (2.11), (2.12))

$$\|\widehat{c}_{N_0,M} - c\| \le \alpha C \sum_{n=1}^{N_0} \left(\left| \widehat{R}_M(u_n) - R(u_n) \right| + \left| \widehat{R}_M(0) - R(0) \right| \right)$$

Hence

$$P\{\|\widehat{c}_{N_0,M} - c\| > \varepsilon\}$$

$$\le \sum_{n=1}^{N_0} P\left\{ \left| \widehat{R}_M(u_n) - R(u_n) \right| > \varepsilon/2N_0\alpha C \right\}$$

$$+ N_0 P\left\{ \left| \widehat{R}_M(0) - R(0) \right| > \varepsilon/2N_0\alpha C \right\} \tag{A.7}$$

for each $\varepsilon > 0$. Since N_0 is finite and by assumption $\left| \widehat{R}_M(u) - R(u) \right| \to 0$ in probability as $M \to \infty$ for each $u \in \{0, u_n; n = 1, 2, ..., N_0\}$, we get the conclusion.   ∎

A.5   Proof of Theorem 2.2

Proof. Denote $\varepsilon = \delta M^{-\tau}/|d_M|$, any $\delta > 0$. From (A.7) in Appendix A.4 we obtain

$$P\left\{ |d_M| \, \|\widehat{c}_{N_0,M} - c\| \, /M^{-\tau} > \delta \right\}$$

$$\le \sum_{n=1}^{N_0} P\left\{ |d_M| \left| \widehat{R}_M(u_n) - R(u_n) \right| /M^{-\tau} > \delta/2N_0\alpha C \right\}$$

$$+ N_0 P\left\{ |d_M| \left| \widehat{R}_M(0) - R(0) \right| /M^{-\tau} > \delta/2N_0\alpha C \right\}$$

which gives the conclusion.   ∎

A.6   Proof of Theorem 2.3

Proof. For each vector norm $\|\cdot\|$ we have

$$\|\widehat{\gamma}_{N,M} - \gamma\| \le \|\widehat{\gamma}_{N,M} - \widehat{\gamma}_N\| + \|\widehat{\gamma}_N - \gamma\| \tag{A.8}$$

Since ([128]) $\|\widehat{\gamma}_N - \gamma\| \to 0$ in probability as $N \to \infty$, it remains to show that $\|\widehat{\gamma}_{N,M} - \widehat{\gamma}_N\| \to 0$ in probability as $N, M \to \infty$, provided that $NM^{-\tau} \to 0$. Owing to equivalence of norms, we further use the 1-vector norm $\|\cdot\|_1$. Let

$$\xi_{N,M} \triangleq \left\| \frac{1}{N}\widehat{\Theta}_{N,M}^T \widehat{\Theta}_{N,M} - \frac{1}{N}\Theta_N^T \Theta_N \right\|_{1,1}$$

$$\varkappa_{N,M} \triangleq \left\| \left(\frac{1}{N}\widehat{\Theta}_{N,M}^T \widehat{\Theta}_{N,M} \right)^{-1} - \left(\frac{1}{N}\Theta_N^T \Theta_N \right)^{-1} \right\|_{1,1}$$

$$\chi_{N,M} \triangleq \left\| \frac{1}{N}\widehat{\Theta}_{N,M}^T Y_N - \frac{1}{N}\Theta_N^T Y_N \right\|_1$$

and moreover $A_N = (\frac{1}{N}\Theta_N^T \Theta_N)^{-1}$, $A = (E\vartheta_1\vartheta_1^T)^{-1}$, $\varepsilon_N = 1/\|A_N\|_{1,1}$, $r_{N,M} = \xi_{N,M}/(\varepsilon_N(\varepsilon_N - \xi_{N,M}))$ where $\|\cdot\|_{1,1}$ is the matrix norm induced by $\|\cdot\|_1$, and let ϑ_{nl} and $\widehat{\vartheta}_{nl,M}$ $(n = 1, 2, ..., N; l = 1, 2, ..., s+1)$ be the elements of Θ_N, $\widehat{\Theta}_{N,M} \in R^{N \times (s+1)}$. Since $\vartheta_{nl} = w_{(n+s)-(l-1)}$, $\widehat{\vartheta}_{nl,M} = \widehat{w}_{(n+s)-(l-1),M}$ and

$$\left| \widehat{\vartheta}_{nl,M} - \vartheta_{nl} \right| \tag{A.9}$$

$$\leqslant \left| \widehat{R}_M(u_{(n+s)-(l-1)}) - R(u_{(n+s)-(l-1)}) \right| + \left| \widehat{R}_M(0) - R(0) \right|,$$

thus, owing to (2.32) and using the Banach inverse map theorem we get for M large

$$\varkappa_{N,M} \leqslant r_{N,M} \tag{A.10}$$

In turn, because the w_n's are bounded, $|w_n| \leqslant a$, some $a > 0$, by the ergodic law of large numbers (e.g. Lemma B.1 in [128]) $\varepsilon_N \to \bar{a}$ in probability as $N \to \infty$, where $\bar{a} = 1/\|A\|_{1,1}$ (see (2.5) and Assumptions 2.1 and 2.3). Further, after standard calculations we obtain

$$\xi_{N,M} \leqslant (s+1)\max_{n,l}\left| \widehat{\vartheta}_{nl,M} - \vartheta_{nl} \right|^2 + 2a(s+1)\max_{n,l}\left| \widehat{\vartheta}_{nl,M} - \vartheta_{nl} \right|$$

$$\chi_{N,M} \leqslant b(s+1)\max_{n,l}\left| \widehat{\vartheta}_{nl,M} - \vartheta_{nl} \right|$$

as $|y_{n+s}| \leqslant b$, some $b > 0$ (see (2.5) and Assumptions 2.1-2.4). This yields respectively

$$P\{\xi_{N,M} > \varepsilon\} \leqslant \sum_{n,l} P\left\{ \left| \widehat{\vartheta}_{nl,M} - \vartheta_{nl} \right| > \sqrt{\frac{\varepsilon}{2(s+1)}} \right\} \tag{A.11}$$

$$+ \sum_{n,l} P\left\{ \left| \widehat{\vartheta}_{nl,M} - \vartheta_{nl} \right| > \frac{\varepsilon}{4a(s+1)} \right\}$$

$$P(\chi_{N,M} > \varepsilon) \leqslant \sum_{n,l} P\left\{ \left| \widehat{\vartheta}_{nl,M} - \vartheta_{nl} \right| > \varepsilon/(b(s+1)) \right\} \tag{A.12}$$

Using Markov's inequality, including (A.9) and the fact that (by boundedness of $\widehat{R}_M$) for $u \in \{0, u_{(n+s)-(l-1)}; n = 1, 2, ..., N; l = 1, 2, ..., s + 1\}$ the rate (2.32) implies asymptotically $E\left|\widehat{R}_M(u) - R(u)\right| = O(M^{-\tau})$, we observe that the r.h.s. of (A.11) and (A.12) is bounded by (A.13) and (A.14) below

$$C(s+1)^2(\sqrt{2/(\varepsilon(s+1))} + 4a/\varepsilon)NM^{-\tau} \tag{A.13}$$
$$C(s+1)^2(b/\varepsilon)NM^{-\tau} \tag{A.14}$$

some $0 < C < \infty$. These two expressions tend to zero as $N, M \to \infty$ for each $\varepsilon > 0$, provided that $NM^{-\tau} \to 0$. The latter along with (A.10) concludes the proof (see (2.8) and (2.14)). ∎

A.7 Proof of Theorem 2.4

Proof. It holds that (here $\|\cdot\|$ means the 1-vector norm $\|\cdot\|_1$ or, respectively, the induced matrix norm $\|\cdot\|_{1,1}$ as in Appendix C)

$$\|\widehat{\gamma}_{N,M} - \widehat{\gamma}_N\| \leqslant \varkappa_{N,M}\chi_{N,M} + \varkappa_{N,M}\|b_N - b\| \tag{A.15}$$
$$+ \chi_{N,M}\|A_N - A\| + d_1\varkappa_{N,M} + d_2\chi_{N,M},$$

where $\varkappa_{N,M}$, $\chi_{N,M}$, A_N, A are as in Appendix C and $b_N = \frac{1}{N}\Theta_N^T Y_N$, $b = E\vartheta_1 y_1$, $d_1 = \|b\|$, $d_2 = \|A\|$. Owing to (A.11)–(A.14), for $M \sim N^{(1+\alpha)/\tau}$, $\alpha > 0$, we get

$$\xi_{N,M} = O(N^{-\alpha}) \text{ in probability,} \tag{A.16}$$
$$\chi_{N,M} = O(N^{-\alpha}) \text{ in probability.} \tag{A.17}$$

Including that $r_{N,M} = \xi_{N,M}/(\varepsilon_N(\varepsilon_N - \xi_{N,M}))$ (see Appendix A.6) and using (A.16), the rate $\varepsilon_N = \bar{a} + O(N^{-1/2})$ in probability and Lemma B.1 and B.2 in Appendix B.3, we obtain $r_{N,M} = O(N^{-\min(1/2,\alpha)})$ in probability. In view of (A.10) this yields

$$\varkappa_{N,M} = O(N^{-\min(1/2,\alpha)}) \text{ in probability.} \tag{A.18}$$

Now, considering that $\|A_N - A\| = O(N^{-1/2})$ in probability, $\|b_N - b\| = O(N^{-1/2})$ in probability along with (A.17) and (A.18), and taking into account Lemma B.1 in Appendix B.3 we conclude that (see (A.15))

$$\|\widehat{\gamma}_{N,M} - \widehat{\gamma}_N\| = O(N^{-\min(1/2,\alpha)}) \text{ in probability.} \tag{A.19}$$

Since $\|\widehat{\gamma}_N - \gamma\| = O(N^{-1/2})$ in probability, thus owing to (A.43), (A.19) and using again Lemma B.1 we obtain eventually $\|\widehat{\gamma}_{N,M} - \gamma\| = O(N^{-\min(1/2,\alpha)})$ in probability. Due to equivalence of norms, the conclusion holds for each norm $\|\cdot\|$. ∎

A.8   Proof of Theorem 2.5

Proof. *(a')* Denoting $A_{N,M} = \widehat{\Psi}_{N,M} - \Psi_N$ and $B_{N,M} = \widehat{\Theta}_{N,M} - \Theta_N$ we have

$$\frac{1}{N}\widehat{\Psi}_{N,M}^T \widehat{\Theta}_{N,M} = \frac{1}{N}(\Psi_N + A_{N,M})^T(\Theta_N + B_{N,M}) =$$
$$= \frac{1}{N}\Psi_N^T \Theta_N + \frac{1}{N}A_{N,M}^T \Theta_N + \tag{A.20}$$
$$+ \frac{1}{N}\Psi_N^T B_{N,M} + \frac{1}{N}A_{N,M}^T B_{N,M}.$$

Since in the case under consideration $\frac{1}{N}\Psi_N^T \Theta_N$ fulfils condition *(a)*, to prove *(a')* it suffices to show that, for example, the $\|\cdot\|_{1,1}$ norms of the remaining three components in (A.20), i.e.

$$\xi_{N,M} = \left\|\frac{1}{N}A_{N,M}^T \Theta_N\right\|_{1,1},$$

$$\chi_{N,M} = \left\|\frac{1}{N}\Psi_N^T B_{N,M}\right\|_{1,1},$$

$$\varkappa_{N,M} = \left\|\frac{1}{N}A_{N,M}^T B_{N,M}\right\|_{1,1},$$

tend to zero in probability as $N, M \to \infty$ and $NM^{-\tau} \to 0$ where $\|\cdot\|_{1,1}$ is the matrix norm induced by the 1-vector norm $\|x\|_1 = \sum_{i=1}^{\dim x}|x[i]|$. This norm can be used in the proof because in the finite dimensional real space all vector norms are equivalent. To show such a property for $\xi_{N,M}$ let us notice that under assumptions as in Section 2.4.4 all elements

$$\vartheta_{kl} = \begin{cases} w_{k-l+1}, & \text{for } l \leqslant s+1 \\ y_{k-l+s+1}, & \text{for } l > s+1 \end{cases}$$

of the matrix Θ_N are bounded in the absolute value, i.e. there exists $0 < b < \infty$ such that $|\vartheta_{kl}| < b$ for each $k = 1, 2, ..., N$ and $l = 1, 2, ..., s+p+1$. Since further

$$\frac{1}{N}A_{N,M}^T \Theta_N = \frac{1}{N}\sum_{k=1}^{N}\begin{bmatrix} \widehat{w}_{k,M} - w_k \\ \widehat{w}_{k-1,M} - w_{k-1} \\ ... \\ \widehat{w}_{k-s-p,M} - w_{k-s-p} \end{bmatrix}[\vartheta_{k1}, \vartheta_{k2}, ..., \vartheta_{k,s+p+1}]$$

we obtain

$$\xi_{N,M} \leqslant b(s+p+1)\max_k|\widehat{w}_{k,M} - w_k|, \qquad 1-(s+p) \leqslant k \leqslant N,$$

and hence

$$P(\xi_{N,M} > \varepsilon) \leqslant \sum_{k=1-(s+p)}^{N} P\left(|\widehat{w}_{k,M} - w_k| > \frac{\varepsilon}{b(s+p+1)}\right).$$

Since by assumption $\widehat{w}_{k,M}$'s are bounded and the rate of convergence of nonparametric estimate is given by (2.32), it holds that

$$E|\widehat{w}_{k,M} - w_k| = O(M^{-\tau}), \ k = 1 - (s+p), ..., N.$$

Hence, using Markov's inequality we have that

$$P(\xi_{N,M} > \varepsilon) \leqslant \frac{C(s+p+1)}{\varepsilon}\left[NM^{-\tau} + (s+p)M^{-\tau}\right], \qquad (A.21)$$

where C is a constant. The right-hand side in (A.21) tends to zero as $N, M \to \infty$ provided that $NM^{-\tau} \to 0$. The convergence of $\chi_{N,M} \to 0$ and $\varkappa_{N,M} \to 0$ in probability can be proved in the same manner.

(b') Including that in the case considered $\frac{1}{N}\Psi_N^T Z_N$ fulfils condition (b) and that

$$\frac{1}{N}\widehat{\Psi}_{N,M}^T Z_N = \frac{1}{N}\Psi_N^T Z_N + \frac{1}{N}A_{N,M}^T Z_N,$$

with $A_{N,M}$ as in part (a') above, the proof of (b') can be obtained by exploiting boundedness of $\bar{z}_k$ (i.e. that $|\bar{z}_k| \leqslant (1 + |a_1| + |a_2| + ... + |a_p|)z_{\max}$; cf. Assumption 2.9) and following the steps of the proof in part (a') with respect to $\varrho_{N,M} = \left\|\frac{1}{N}A_{N,M}^T Z_N\right\|_1$. ∎

A.9   Proof of Theorem 2.7

Proof. Let us put, for convenience, $\bar{z}_{\max} = 1$. Taking account of (2.35) one can write

$$\left\|\Delta_N^{(IV)}(\Psi_N)\right\|_2^2 = \Delta_N^{(IV)T}(\Psi_N)\Delta_N^{(IV)}(\Psi_N) = Z_N^{*T}\Gamma_N^T\Gamma_N Z_N^*,$$

and hence the quality index (2.36) for N large enough is

$$Q(\Psi_N) = \max_{\|Z_N^*\|_2 \leqslant 1} \langle Z_N^*, \Gamma_N^T\Gamma_N Z_N^*\rangle = \|\Gamma_N\|^2 = \lambda_{\max}\left(\Gamma_N^T\Gamma_N\right),$$

where $\|\cdot\|$ is a spectral matrix-norm induced by the Euclidean vector-norm, and $\lambda_{\max}(A)$ is the greatest absolute eigenvalue of A. Since (see [112], Chapter 1)

$$\lambda_{\max}\left(\Gamma_N^T\Gamma_N\right) = \lambda_{\max}\left(\Gamma_N\Gamma_N^T\right)$$

we conclude that

$$Q(\Psi_N) = \max_{\|\zeta\|_2 \leq 1} \left\langle \zeta, \Gamma_N \Gamma_N^T \zeta \right\rangle$$

$$= \max_{\|\zeta\|_2 \leq 1} \left\langle \zeta, \left(\frac{1}{N}\Psi_N^T \Theta_N\right)^{-1} \left(\frac{1}{N}\Psi_N^T \Psi_N\right) \left(\frac{1}{N}\Theta_N^T \Psi_N\right)^{-1} \zeta \right\rangle,$$

where Ψ_N is an arbitrary instrumental matrix

$$\Psi_N = (\psi_1, \psi_2, ..., \psi_N)^T$$

of the form as in (2.24) fulfilling the conditions *(a)* and *(b)*. Since

$$\Theta_N = (\vartheta_1, \vartheta_2, ..., \vartheta_N)^T,$$

where

$$\vartheta_k = (w_k, w_{k-1}, ..., w_{k-s}, y_{k-1}, y_{k-2}, ..., y_{k-p})^T$$

for $k = 1, 2, ..., N$, and

$$\Psi_N^* = (\psi_1^*, \psi_2^*, ..., \psi_N^*)^T,$$

where

$$\psi_k^* = (w_k, w_{k-1}, ..., w_{k-s}, v_{k-1}, v_{k-2}, ..., v_{k-p})^T$$

for $k = 1, 2, ..., N$, and moreover

$$y_{k-r} = v_{k-r} + z_{k-r}, \qquad r = 1, 2, ..., p$$

one can write in fact

$$\vartheta_k = \psi_k^* + \widetilde{z}_k,$$

where $\widetilde{z}_k$ is the following output noise vector

$$\widetilde{z}_k = (0, 0, ..., 0, z_{k-1}, z_{k-2}, ..., z_{k-p})^T$$

for $k = 1, 2, ..., N$. This yields eventually

$$\Theta_N = \Psi_N^* + \widetilde{Z}_N,$$

where

$$\widetilde{Z}_N = (\widetilde{z}_1, \widetilde{z}_2, ..., \widetilde{z}_N)^T,$$

and leads further to the relation

$$\frac{1}{N}\Psi_N^T \Theta_N = \frac{1}{N}\Psi_N^T \left(\Psi_N^* + \widetilde{Z}_N\right) = \frac{1}{N}\Psi_N^T \Psi_N^* + \frac{1}{N}\Psi_N^T \widetilde{Z}_N,$$

i.e.,

$$\frac{1}{N}\Psi_N^T\Theta_N - \frac{1}{N}\Psi_N^T\Psi_N^* = \frac{1}{N}\sum_{k=1}^{N}\Psi_N^T\widetilde{Z}_N.$$

Owing to denotation (2.25) in Remark 2.5 we see that

$$\widetilde{Z}_N = \left[0,0,...,0; \widetilde{Z}_N^{(1)}, \widetilde{Z}_N^{(2)},..., \widetilde{Z}_N^{(p)}\right]_{N\times[(s+1)+p]},$$

i.e.,

$$\frac{1}{N}\Psi_N^T\widetilde{Z}_N = \left[0,0,...,0; \frac{1}{N}\Psi_N^T\widetilde{Z}_N^{(1)}, \frac{1}{N}\Psi_N^T\widetilde{Z}_N^{(2)},..., \frac{1}{N}\Psi_N^T\widetilde{Z}_N^{(p)}\right],$$

and due to (2.26) in Remark 2.5 it holds that

$$\frac{1}{N}\Psi_N^T\widetilde{Z}_N \to 0 \text{ in probability as } N \to \infty.$$

Hence

$$\frac{1}{N}\Psi_N^T\Theta_N - \frac{1}{N}\Psi_N^T\Psi_N^* \to 0 \text{ in probability as } N \to \infty,$$

which means that for each, arbitrarily small, $\varepsilon > 0$ we have

$$P\left\{\left\|\frac{1}{N}\Psi_N^T\Theta_N - \frac{1}{N}\Psi_N^T\Psi_N^*\right\| \geqslant \varepsilon\right\} \to 0 \text{ as } N \to \infty,$$

or that asymptotically, as $N \to \infty$, it holds that

$$P\left\{\left\|\frac{1}{N}\Psi_N^T\Theta_N - \frac{1}{N}\Psi_N^T\Psi_N^*\right\| \neq 0\right\} \simeq 0,$$

yielding subsequently

$$\frac{1}{N}\Psi_N^T\Theta_N \simeq \frac{1}{N}\Psi_N^T\Psi_N^* \tag{A.22}$$

with probability 1. Since for N large, with probability 1, $\frac{1}{N}\Psi_N^T\Psi_N^*$ can be applied instead of $\frac{1}{N}\Psi_N^T\Theta_N$, we obtain

$$Q(\Psi_N) = \max_{\|\zeta\|_2 \leq 1} \left\langle \zeta, \left(\frac{1}{N}\Psi_N^T\Psi_N^*\right)^{-1}\left(\frac{1}{N}\Psi_N^T\Psi_N\right)\left(\frac{1}{N}\Psi_N^{*T}\Psi_N\right)^{-1}\zeta\right\rangle$$

with probability 1. Making now use of Lemma B.3 in Appendix B.3 for $M_1 = \frac{1}{\sqrt{N}}\Psi_N^*$ and $M_2 = \frac{1}{\sqrt{N}}\Psi_N$, and including (A.22) we get asymptotically

$$\zeta^T\Gamma_N\Gamma_N^T\zeta \geq \zeta^T\left(\frac{1}{N}\Psi_N^{*T}\Psi_N^*\right)^{-1}\zeta$$

with probability 1 for each vector ζ. Therefore

$$Q\left(\Psi_N\right) = \max_{\|\zeta\|_2 \le 1}\left(\zeta^T \Gamma_N \Gamma_N^T \zeta\right) \ge \max_{\|\zeta\|_2 \le 1}\left(\zeta^T \left(\frac{1}{N}\Psi_N^{*T}\Psi_N^*\right)^{-1}\zeta\right). \quad \text{(A.23)}$$

Since for $\Psi_N = \Psi_N^*$ both sides in (A.23) are equal, $Q\left(\Psi_N\right)$ attains minimum as $\Psi_N = \Psi_N^*$. ∎

A.10 Proof of Theorem 2.8

Proof. Owing to (2.45) and (2.51), we see that

$$\widehat{Q}_{N_0,M}(c) = Q_{N_0}(c) + 2\sum_{n=1}^{N_0}(w_n - \mu(\bar{u}_n, c))(\widehat{w}_{n,M} - w_n) + \sum_{n=1}^{N_0}(\widehat{w}_{n,M} - w_n)^2,$$

and hence (by Cauchy inequality $(x+y)^2 \le 2x^2 + 2y^2$) we obtain

$$\left|\widehat{Q}_{N_0,M}(c) - Q_{N_0}(c)\right| \le \left(2\sum_{n=1}^{N_0}|\mu(\bar{u}_n, c^*) - \mu(\bar{u}_n, c)|\right) \cdot \Delta_M(0) +$$

$$+ 2\sum_{n=1}^{N_0}|\mu(\bar{u}_n, c^*) - \mu(\bar{u}_n, c)| \cdot \Delta_M(\bar{u}_n) +$$

$$+ 2N_0\Delta_M^2(0) + 2\sum_{n=1}^{N_0}\Delta_M^2(\bar{u}_n),$$

where $\Delta_M(\bar{u}_n) = \left|\widehat{R}_M(\bar{u}_n) - R(\bar{u}_n)\right|$. In particular, for each $c \in \overline{C}$ (a closure of C) by continuity of $\mu(u, c)$ (Assumption 2.13) we have

$$|\mu(\bar{u}_n, c) - \mu(\bar{u}_n, c^*)| \le A_n,$$

where A_n is a constant (possibly depending on the input point $\bar{u}_n$), and eventually we get

$$\sup_{c \in \overline{C}}\left|\widehat{Q}_{N_0,M}(c) - Q_{N_0}(c)\right| \le \quad \text{(A.24)}$$

$$\le 2A\left[N_0\Delta_M(0) + \sum_{n=1}^{N_0}\Delta_M(\bar{u}_n)\right] + 2\left[N_0\Delta_M^2(0) + \sum_{n=1}^{N_0}\Delta_M^2(\bar{u}_n)\right],$$

where $A = \max\{A_1, A_2, ..., A_{N_0}\}$. Under assumptions of theorem

$$\sup_{c \in \overline{C}}\left|\widehat{Q}_{N_0,M}(c) - Q_{N_0}(c)\right| \to 0 \text{ in probability,}$$

as $M \to \infty$, and thus $\widehat{c}_{N_0,M} \to c^*$ in probability as M grows (cf. [142], [139]). Asymptotically (i.e. for M large) we have (see (2.47))

$$Q_{N_0}(\widehat{c}_{N_0,M}) \geq \delta \left\| \widehat{c}_{N_0,M} - c^* \right\|.$$

and hence, for each $\varepsilon > 0$, it holds that

$$P\left\{ \left\| \widehat{c}_{N_0,M} - c^* \right\| > (\varepsilon/\delta)^{1/2} \right\} \leq P\left\{ Q_{N_0}(\widehat{c}_{N_0,M}) > \varepsilon \right\}.$$

Since further

$$Q_{N_0}(\widehat{c}_{N_0,M}) = Q_{N_0}(\widehat{c}_{N_0,M}) - Q_{N_0}(c^*) \leq 2 \sup_{c \in C} \left| \widehat{Q}_{N_0,M}(c) - Q_{N_0}(c) \right| \leq$$

$$\leq 2 \sup_{c \in \overline{C}} \left| \widehat{Q}_{N_0,M}(c) - Q_{N_0}(c) \right|$$

we conclude (2.53) including (A.24) and (2.52).   ∎

A.11   Proof of Theorem 2.9

Owing to (2.70) and by independence between u_k and d_k the condition *(C2)* is obvious. Since $\overline{u}_k$ is assumed to be a strongly persistently exciting signal, it is sufficient to prove the following lemma.

Lemma A.4. *If* $\det E\overline{u}_k\overline{u}_k^T \neq 0$, *then* $\det E\overline{x}_k\overline{u}_k^T \neq 0$.

Proof. Let us decompose the random process x_k on the white noise u_k and the remaining part r_k

$$x_k = u_k + \sum_{i=1}^{\infty} g_i u_{k-i} = u_k + r_k.$$

According to the Taylor series expansion of x_k^w we have

$$x_k^w = u_k^w + \frac{r_k}{1!} w u_k^{w-i} + \frac{r_k}{1!} w u_k^{w-i} + \frac{r_k^2}{2!} w(w-1) u_k^{w-i} + \ldots$$

$$\ldots + \frac{r_k^w}{w!} w! u_k^0 = u_k^w + \sum_{i=1}^{w} r_{i,k} u_k^{w-i},$$

where

$$r_{i,k} = \frac{r_k^i}{i!} w(w-1) \cdot \ldots \cdot (w-i),$$

which yields to the following relationship

$$\overline{x}_k = R_k \overline{u}_k,$$

between vectors

$$\overline{x}_k = \left(1, x_k, x_k^2, ..., x_k^w, ..., x_k^{W-1}\right)^T,$$

and

$$\overline{u}_k = \left(1, u_k, u_k^2, ..., u_k^w, ..., u_k^{W-1}\right)^T,$$

with the matrix R_k of the form

$$R_k = \begin{bmatrix} 1 & 0 & 0 & 0 & 0 \\ r_{1,k} & 1 & 0 & 0 & 0 \\ r_{2,k} & r_{1,k} & 1 & 0 & 0 \\ \vdots & \vdots & \ddots & \ddots & 0 \\ \cdots & r_{3,k} & r_{2,k} & r_{1,k} & 1 \end{bmatrix}.$$

Since R_k is of full rank and its elements are independent of $\overline{u}_k$, under the assumption that $\det E\overline{u}_k\overline{u}_k^T \neq 0$ we conclude that

$$\det E\overline{x}_k\overline{u}_k^T = \det ER_k \cdot \det E\overline{u}_k\overline{u}_k^T \neq 0,$$

which ends the proof.   ∎

A.12   Proof of Theorem 2.10

Proof. From (2.83) we conclude that for all $w = 1, 2, ..., W$ and $k = 1, 2, ..., N$, it holds that

$$\left|\widehat{u}_{k,M}^w - u_k^w\right| = O(M^{-w\tau}),$$

and consequently

$$\|\psi_{k,M} - \psi_k\| = O(M^{-w\tau}),$$

with probability 1, as $M \to \infty$. Now, Theorem 2.10 may be proved in a similar manner as Theorem 3 in [65], with obvious substitutions, and hence the rest of the proof is here omitted.   ∎

A.13   Proof of Theorem 2.11

Proof. Owing to the fact that $Ed_k = 0$ and by independence between u_k and d_k the condition *(C2)* is obvious. Since $\overline{\psi}_k$ is a strongly persistently exciting signal, it is sufficient to prove the following lemma, which guarantees that the instrumental variables estimate is well defined (i.e., the condition *(C1)*).   ∎

Lemma A.5. *If* $\det E\overline{\psi}_k\overline{\psi}_k^T \neq 0$, *then* $\det E\phi_k\overline{\psi}_k^T \neq 0$.

Proof. Assuming that $\lambda_0 = 1$, let us decompose the process x_k on the white noise $\eta(u_k)$ and the remaining part, i.e.,

$$x_k = \eta(u_k) + \sum_{i=1}^{\infty} \lambda_i \eta(u_{k-i}) + \varepsilon_k = \eta(u_k) + s_k,$$

where $s_k = \sum_{i=1}^{\infty} \lambda_i \eta(u_{k-i}) + \varepsilon_k$ is statistically independent of $\eta(u_k)$. According to the Taylor series expansion of x_k^w we have

$$x_k^w = \eta^w(u_k) + \frac{s_k}{1!} w \eta^{w-1}(u_k) + \frac{s_k^2}{2!} w(w-1) \eta^{w-2}(u_k) + \dots$$

$$\dots + \frac{r_k^w}{w!} w! \eta^0(u_k) = \eta^w(u_k) + \sum_{i=1}^{w} s_{i,k} \eta^{w-i}(u_k),$$

where

$$s_{i,k} = \frac{s_k^i}{i!} w(w-1) \cdot \dots \cdot (w-i),$$

which yields to the following relationship between vectors

$$\phi_k = \left(1, x_k, x_k^2, \dots, x_k^w, \dots, x_k^{W-1}\right)^T$$

and

$$\overline{\psi}_k = \left(1, u_k, u_k^2, \dots, u_k^w, \dots, u_k^{W-1}\right)^T$$

$$\overline{\psi}_k = S_k \phi_k,$$

with the matrix S_k of the form

$$S_k = \begin{bmatrix} 1 & 0 & 0 & 0 & 0 \\ s_{1,k} & 1 & 0 & 0 & 0 \\ s_{2,k} & s_{1,k} & 1 & 0 & 0 \\ \vdots & \vdots & \ddots & \ddots & 0 \\ \cdots & s_{3,k} & s_{2,k} & s_{1,k} & 1 \end{bmatrix}.$$

Since S_k is of full rank and its elements are independent of $\overline{\psi}_k$, under the assumption that $\det E\overline{\psi}_k \overline{\psi}_k^T \neq 0$ we conclude that

$$\det E\phi_k \overline{\psi}_k^T = \det ES_k \cdot \det E\overline{\psi}_k \overline{\psi}_k^T \neq 0,$$

which ends the proof.   ∎

A.14 Proof of Theorem 2.13

Proof. It is obvious that the system with the input u_k and the filtered output y_k^f also belongs to the class of Hammerstein systems, and has the same static characteristic $\mu(u)$. To prove (2.131) it remains to show that the resulting linear dynamics (i.e. $\{\gamma_i\}_{i=0}^{\infty}$ in a cascade with the filter (2.128)) and the resulting output noise (i.e. $\{z_k\}$ transferred as in (2.128)) fulfill Assumptions 2.29 and 2.30 as $N \to \infty$. Let us emphasize that the parameters of the filter (2.128) are random. The estimation error in Step 1 has the form

$$\Delta_N \triangleq \widehat{\theta}_N - \theta = \left(\frac{1}{N} \Psi_N^T \Phi_N \right)^{-1} \left(\frac{1}{N} \Psi_N^T D_N \right),$$

and under ergodicity of the processes $\{m(u_k)\}$, $\{y_k\}$ and $\{d_k\}$ it holds that

$$\frac{1}{N} \Psi_N^T \Phi_N = \frac{1}{N} \sum_{k=1}^{N} \phi_k \psi_k^T \to E\phi_k \psi_k^T,$$

and

$$\frac{1}{N} \Psi_N^T D_N = \frac{1}{N} \sum_{k=1}^{N} d_k \psi_k^T \to Ed_k \psi_k^T,$$

with probability 1, as $N \to \infty$. Since $\{d_k\}$ is zero-mean and d_{k_1} is independent of u_{k_2} for any time instants k_1 and k_2 such that $k_1 > k_2$, it holds that

$$Ed_k \psi_k^T = E\left(d_k \cdot (m(u_{k-1}), m(u_{k-2}), ..., m(u_{k-p}), 1)\right) = 0.$$

Using the results presented in [49], concerning dependence between input-output cross-correlation in Hammerstein system and the terms of impulse response we have

$$\sigma_{m(u)} \sigma_y \gamma_{|i-j|} = E\{(y_{k-i} - Ey)(m(u_{k-j}) - Em(u))\} =$$
$$= E\{y_{k-i}(m(u_{k-j}) - Em(u))\} - Ey \cdot E(m(u_{k-j}) - Em(u)) =$$
$$= Ey_{k-i}m(u_{k-j}) - Ey \cdot Em(u),$$

where $\sigma_{m(u)}^2 = \mathrm{var}\, m(u)$, and hence

$$E\phi_k \psi_k^T = \begin{bmatrix} \Gamma\, \overline{y} \\ I \end{bmatrix},$$

where $\overline{y} = (Ey, Ey, ..., Ey)^T$, $I = (0, 0, ..., 0, 1)$, and

$$\Gamma = \begin{bmatrix} \overline{\gamma}_0 & \overline{\gamma}_1 & \cdots & \overline{\gamma}_{p-1} \\ \overline{\gamma}_1 & \overline{\gamma}_0 & \ddots & \vdots \\ \vdots & \ddots & \ddots & \overline{\gamma}_1 \\ \overline{\gamma}_{p-1} & \cdots & \overline{\gamma}_1 & \overline{\gamma}_0 \end{bmatrix},$$

where $\overline{\gamma}_l = m\sigma_{m(u)}\sigma_y \gamma_l + Ey \cdot Em(u)$. For AR($p$) process $\{y_k\}$ the matrix Γ is of full rank (see [15], [48]). Since $\det E\phi_k \psi_k^T \neq 0$, the estimate (2.126) is well defined with probability 1, as $N \to \infty$. Consequently, for $i = 1, 2, ..., p$ it holds that

$$MSE\left[\widehat{a}_{i,N}\right] \triangleq E\left(\widehat{a}_{i,N} - a_i\right)^2 = O\left(\frac{1}{N}\right),$$

and

$$\left|\sum_{i=1}^{p}(a_i - \widehat{a}_{i,N})y_{k-i}\right| = O\left(\frac{1}{\sqrt{N}}\right)$$

in probability, as $N \to \infty$. Inserting (2.122) to (2.128) we get

$$y_k^f = w_k + \widetilde{z}_k + \sum_{i=1}^{p}(a_i - \widehat{a}_{i,N})y_{k-i}. \qquad (A.25)$$

Since for $N \to \infty$ it holds that $|\sum_{i=1}^{p}(a_i - \widehat{a}_{i,N})y_{k-i}| = 0$ with probability 1, we obtain $y_k^f = w_k + \widetilde{z}_k$ with probability 1, as $N \to \infty$, which guarantees fulfillment of Assumptions 2.29 and 2.30.   ∎

A.15   Proof of Proposition 3.1

Proof. The condition (3.12) is fulfilled with probability 1 for each $j > j_0$, where $j_0 = \lfloor \log_\lambda d \rfloor$ is the solution of the following inequality

$$\frac{d}{\lambda^j} \geqslant 2u_{\max} = 1.$$

On the basis of Assumption 3.2, analogously as in (3.9), we obtain

$$|x_k - x| \leqslant \sum_{j=0}^{j_0} \lambda^j \frac{d}{\lambda^j} + \frac{\lambda^{j_0+1}}{1-\lambda} = d\left(j_0 + 1 + \frac{\lambda}{1-\lambda}\right),$$

which yields (3.13).   ∎

A.16   Proof of Theorem 3.3

Proof. Let us denote the probability of selection as

$$p(M) \triangleq P\left(\Delta_k(x) \leqslant h(M)\right).$$

To prove (3.16) it suffices to show that (see (19) and (22) in [90])

$$h(M) \to 0, \qquad (A.26)$$
$$Mp(M) \to \infty, \qquad (A.27)$$

as $M \to \infty$. They assure vanishing of the bias and variance of $\widehat{\mu}_M(x)$, respectively. Since under assumptions of Theorem 3.3

$$d(M) \to 0 \Rightarrow h(M) \to 0, \tag{A.28}$$

in view of (3.13), the bias-condition (A.26) is obvious. For the variance-condition (A.27) we have that

$$p(M) \geqslant P \left\{ \bigcap_{j=0}^{\min(k,j_0)} \left(|u_{k-j} - x| < \frac{d(M)}{\lambda^j} \right) \right\} \geqslant$$

$$\geqslant P \left\{ \bigcap_{j=0}^{\min(k,j_0)} \left(|u_{k-j} - x| < \frac{d(M)}{\lambda^j} \right) \right\} =$$

$$= \prod_{j=0}^{j_0} P \left(|u_{k-j} - x| < \frac{d(M)}{\lambda^j} \right) \geqslant$$

$$\geqslant \varepsilon \frac{d(M)}{\lambda^0} \cdot \varepsilon \frac{d(M)}{\lambda^1} \cdot \dots \cdot \varepsilon \frac{d(M)}{\lambda^{j_0}} =$$

$$= \frac{(\varepsilon d(M))^{j_0+1}}{\lambda^{\frac{j_0(j_0+1)}{2}}} = \left(\frac{\varepsilon d(M)}{\lambda^{\frac{j_0}{2}}} \right)^{j_0+1} =$$

$$= \left(\varepsilon \sqrt{d(M)} \right)^{j_0+1} = \varepsilon \cdot d(M)^{\frac{1}{2} \log_\lambda d(M) + \log_\lambda \varepsilon + \frac{1}{2}}. \tag{A.29}$$

By inserting $d(M) = M^{-\gamma(M)} = (1/\lambda)^{-\gamma(M) \log_{1/\lambda} M}$ to (A.29) we obtain

$$M \cdot p(M) = \varepsilon \cdot M^{1-\gamma(M)\left(\frac{1}{2}\gamma(M) \log_{1/\lambda} M + \log_\lambda \varepsilon + \frac{1}{2} \right)}. \tag{A.30}$$

For $\gamma(M) = \left(\log_{1/\lambda} M \right)^{-w}$ and $w \in (\frac{1}{2}, 1)$ from (A.30) we simply conclude (A.27) and consequently (3.16). ∎

A.17 Proof of Theorem 4.1

Proof. For a given estimation point x of $\mu(\cdot)$ define a "weighted distance" $\delta_k(x)$ between input measurements $u_k, u_{k-1}, u_{k-2}, \dots, u_1$ and x, similarly as for Wiener system (see (3.7) and (3.8)). Making use of Assumptions 4.5 and 4.2 we obtain that

$$|x_k - x| = \left| \sum_{j=0}^{\infty} \lambda_j u_{k-j} - \sum_{j=0}^{\infty} \lambda_j x \right| =$$

$$= \left| \sum_{j=0}^{\infty} \lambda_j \left(u_{k-j} - x \right) \right| =$$

$$= \left| \sum_{j=0}^{k-1} \lambda_j \left(u_{k-j} - x \right) + \sum_{j=k}^{\infty} \lambda_j \left(u_{k-j} - x \right) \right| \leqslant$$

$$\leqslant \sum_{j=0}^{k-1} |\lambda_j| \, |u_{k-j} - x| + 2 u_{\max} \sum_{j=k}^{\infty} |\lambda_j| \leqslant$$

$$\leqslant \delta_k(x) + \frac{\lambda^k}{1-\lambda} \triangleq \Delta_k(x). \tag{A.31}$$

Observe that if in turn

$$\Delta_k(x) \leqslant h(N), \tag{A.32}$$

then the true (but unknown) interaction input x_k is located close to x, provided that $h(N)$ (further, a calibration parameter) is small. If, for each $j = 0, 1, ..., \infty$ and some $d > 0$, it holds that

$$|u_{k-j} - x| \leqslant \frac{d}{\lambda^j}, \tag{A.33}$$

then

$$|x_k - x| \leqslant d \log_\lambda d + d \frac{1}{1-\lambda}. \tag{A.34}$$

The condition (A.33) is fulfilled with probability 1 for each $j > j_0$, where $j_0 = \lfloor \log_\lambda d \rfloor$ is the solution of the following inequality

$$\frac{d}{\lambda^j} \geqslant 2 u_{\max} = 1.$$

On the basis of Assumption 4.2, analogously as in (A.31), we obtain that

$$|x_k - x| \leqslant \sum_{j=0}^{j_0} \lambda^j \frac{d}{\lambda^j} + \frac{\lambda^{j_0+1}}{1-\lambda} = d \left(j_0 + 1 + \frac{\lambda}{1-\lambda} \right),$$

which yields (A.34). For the Wiener-Hammerstein (sandwich) system we have

$$\left| \bar{y}_k - \mu(x) \right| = \left| \sum_{i=0}^{\infty} \gamma_i \mu \left(x_{k-i} \right) - \sum_{i=0}^{\infty} \gamma_i \mu \left(x \right) \right| =$$

$$= \left| \sum_{i=0}^{\infty} \gamma_i \mu \left(\sum_{j=0}^{\infty} \lambda_j u_{k-i-j} \right) - \sum_{i=0}^{\infty} \gamma_i \mu \left(\sum_{j=0}^{\infty} \lambda_j x \right) \right| =$$

$$= \left| \sum_{i=0}^{\infty} \gamma_i \left[\mu \left(\sum_{j=0}^{\infty} \lambda_j u_{k-i-j} \right) - \mu \left(\sum_{j=0}^{\infty} \lambda_j x \right) \right] \right| \leqslant$$

$$\leqslant l \sum_{i=0}^{\infty} |\gamma_i| \left| \sum_{j=0}^{\infty} \lambda_j \left(u_{k-i-j} - x \right) \right| \leqslant$$

$$\leqslant l \sum_{i=0}^{\infty} |\gamma_i| \sum_{j=0}^{\infty} |\lambda_j| \left| u_{k-i-j} - x \right| = l \sum_{i=0}^{\infty} \varkappa_i \left| u_{k-i} - x \right| \qquad (A.35)$$

where the sequence $\{\varkappa_i\}_{i=0}^{\infty}$ obviously fulfills the condition $|\varkappa_i| \leqslant \lambda^i$. Let us denote the probability of selection as $p(N) \triangleq P\left(\Delta_k(x) \leqslant h(N) \right)$. To prove (4.8) it suffices to show that (see (19) and (22) in [90])

$$h(N) \to 0, \qquad (A.36)$$

$$Np(N) \to \infty, \qquad (A.37)$$

as $N \to \infty$. The conditions (A.36) and (A.37) assure vanishing of the bias and variance of $\widehat{\mu}_N(x)$, respectively. Since under assumptions of Theorem 4.1

$$d(N) \to 0 \Rightarrow h(N) \to 0, \qquad (A.38)$$

in view of (A.34), the bias-condition (A.36) is obvious. For the variance-condition (A.37) we have

$$p(N) \geqslant P \left\{ \bigcap_{j=0}^{\min(k,j_0)} \left(|u_{k-j} - x| < \frac{d(N)}{\lambda^j} \right) \right\} \geqslant$$

$$\geqslant P \left\{ \bigcap_{j=0}^{\min(k,j_0)} \left(|u_{k-j} - x| < \frac{d(N)}{\lambda^j} \right) \right\} =$$

$$= \prod_{j=0}^{j_0} P \left(|u_{k-j} - x| < \frac{d(N)}{\lambda^j} \right) \geqslant \qquad (A.39)$$

$$\geqslant \varepsilon \frac{d(N)}{\lambda^0} \cdot \varepsilon \frac{d(N)}{\lambda^1} \cdot \ldots \cdot \varepsilon \frac{d(N)}{\lambda^{j_0}} =$$

$$= \frac{(\varepsilon d(N))^{j_0+1}}{\lambda^{\frac{j_0(j_0+1)}{2}}} = \left(\frac{\varepsilon d(N)}{\lambda^{\frac{j_0}{2}}} \right)^{j_0+1} =$$

$$= \left(\varepsilon \sqrt{d(N)} \right)^{j_0+1} = \varepsilon \cdot d(N)^{\frac{1}{2} \log_\lambda d(N) + \log_\lambda \varepsilon + \frac{1}{2}}.$$

By inserting $d(N) = N^{-\gamma(N)} = (1/\lambda)^{-\gamma(N) \log_{1/\lambda} N}$ to (A.39) we obtain

$$N \cdot p(N) = \varepsilon \cdot N^{1 - \gamma(N) \left(\frac{1}{2} \gamma(N) \log_{1/\lambda} N + \log_\lambda \varepsilon + \frac{1}{2} \right)}. \qquad (A.40)$$

For $\gamma(N) = \left(\log_{1/\lambda} N \right)^{-w}$ and $w \in \left(\frac{1}{2}, 1 \right)$ from (A.40) we simply conclude (A.37) and consequently (4.8). ∎

A.18 The Necessary Condition for the 3-Stage Algorithm

Lemma A.6. *If* $\det(B^T A) \neq 0$, *for a given matrices* $A, B \in R^{\alpha \times \beta}$ *with finite elements, then* $\det(A^T A) \neq 0$.

Proof. Let $\det(A^T A) = 0$, i.e., $rank(A^T A) < \beta$. From the obvious property that

$$rank(A^T A) = rank(A)$$

we conclude that one can find the non-zero vector $\xi \in R^\beta$, such that $A\xi = 0$. Multiplying this equation by B^T we get $B^T A\xi = 0$, and hence $\det(B^T A) = 0$. For $A = \frac{1}{\sqrt{N}} \Phi_N$ and $B = \frac{1}{\sqrt{N}} \Psi_N$ we conclude that the necessary condition for $\frac{1}{N} \Psi_N^T \Phi_N$ to be of full rank is $\det(\frac{1}{N} \Phi_N^T \Phi_N) \neq 0$, i.e., persistent excitation of $\{\phi_k\}$. ∎

A.19   Proof of Theorem 5.1

Proof. From the Slutzky theorem (cf. [24] and Appendix B.5) we have

$$\text{Plim}_{N\to\infty}(\Delta_N^{(IV)}) = \left(\text{Plim}_{N\to\infty}\left(\frac{1}{N}\Psi_N^T\Phi_N\right)\right)^{-1}\text{Plim}_{N\to\infty}\left(\frac{1}{N}\Psi_N^T Z_N\right),$$

and directly from the conditions (C2) and (C3) it holds that

$$P\lim_{N\to\infty}\left(\Delta_N^{(IV)}\right) = 0. \tag{A.41}$$

∎

A.20   Proof of Theorem 5.2

Proof. Let us define the scalar random variable

$$\xi_N = \left\|\Delta_N^{(IV)}\right\| = \left\|\widehat{\theta}_N^{(IV)} - \theta\right\|$$

where $\|\ \|$ denotes any vector norm. We must show that

$$P\left\{r_N\frac{\xi_N}{a_N} > \varepsilon\right\} \to 0, \text{ as } N \to \infty,$$

for each $\varepsilon > 0$, each $r_N \to 0$ and $a_N = \frac{1}{\sqrt{N}}$ (see Definition B.4 on page 224). Using Lemma B.6 (page 224), to prove that $\xi_N = O(\frac{1}{\sqrt{N}})$ in probability, we show that $\xi_N = O(\frac{1}{N})$ in the mean square sense. Introducing

$$A_N = \frac{1}{N}\Psi_N^T\Phi_N = \frac{1}{N}\sum_{k=1}^{N}\psi_k\phi_k^T,$$

$$B_N = \frac{1}{N}\Psi_N^T Z_N = \frac{1}{N}\sum_{k=1}^{N}\psi_k z_k,$$

we obtain that

$$\Delta_N^{(IV)} = A_N^{-1}B_N. \tag{A.42}$$

Under Assumptions 5.3–5.6 (page 114) we conclude that the system output y_k is bounded, i.e., $|y_k| < y_{max} < \infty$. Moreover, under condition (C1) (page 119), it holds that

$$\left|A_N^{i,j}\right| \leq \psi_{max}p_{max} < \infty,$$

for $j = 1, 2, ..., m(n+1)$, and

$$\left| A_N^{i,j} \right| \leq \psi_{\max} p_{\max} < \infty,$$

for $j = m(n+1) + 1, ..., m(n+1) + pq$, so each element of A_N is bounded. Similarly, one can show boundedness of elements of the vector B_N. The norm of the error error $\Delta_N^{(IV)}$ given by (A.42) can be evaluated as follows

$$\xi_N = \left\| \Delta_N^{(IV)} \right\| = \left\| \left(\frac{1}{N} \Psi_N^T \Phi_N \right)^{-1} \left(\frac{1}{N} \Psi_N^T Z_N \right) \right\| \leq$$

$$\leq \left\| \left(\frac{1}{N} \Psi_N^T \Phi_N \right)^{-1} \right\| \left\| \frac{1}{N} \Psi_N^T Z_N \right\| \leq$$

$$\leq c \left\| \frac{1}{N} \Psi_N^T Z_N \right\| = c \left\| \frac{1}{N} \sum_{k=1}^{N} \psi_k z_k \right\|$$

where c is some positive constant. Obviously, one can find $\alpha \geq 0$ such that

$$c \left\| \frac{1}{N} \sum_{k=1}^{N} \psi_k z_k \right\| \leq \alpha c \sum_{i=1}^{\dim \psi_k} \left(\frac{1}{N} \left| \sum_{k=1}^{N} \psi_{k,i} z_k \right| \right).$$

and hence

$$\xi_N^2 = \left\| \Delta_N^{(IV)} \right\|^2 \leq \alpha^2 c^2 \left[\sum_{i=1}^{\dim \psi_k} \left(\frac{1}{N} \left| \sum_{k=1}^{N} \psi_{k,i} z_k \right| \right) \right]^2 \leq$$

$$\leq \alpha^2 c^2 \dim \psi_k \sum_{i=1}^{\dim \psi_k} \left(\frac{1}{N} \left| \sum_{k=1}^{N} \psi_{k,i} z_k \right| \right)^2 =$$

$$= \alpha^2 c^2 \dim \psi_k \sum_{i=1}^{\dim \psi_k} \frac{1}{N^2} \left(\sum_{k=1}^{N} \psi_{k,i} z_k \right)^2.$$

Moreover, for uncorrelated processes $\{\psi_k\}$ and $\{z_k\}$ (see condition (C3) on page 119) we have

$$E\xi_N^2 \le \alpha^2 c^2 \dim\psi_k \sum_{i=1}^{\dim\psi_k} \frac{1}{N^2} E\left(\sum_{k=1}^{N} \psi_{k,i} z_k\right)^2 =$$

$$= \alpha^2 c^2 \dim\psi_k \sum_{i=1}^{\dim\psi_k} \frac{1}{N^2} E\left[\sum_{k_1=1}^{N}\sum_{k_2=1}^{N} \psi_{k_1,i}\psi_{k_2,i} z_{k_1} z_{k_2}\right] \le$$

$$\le \alpha^2 c^2 \dim\psi_k \sum_{i=1}^{\dim\psi_k} \frac{1}{N^2} \sum_{k_1=1}^{N}\sum_{k_2=1}^{N} |E\left[\psi_{k_1,i}\psi_{k_2,i}\right]| \, |E\left[z_{k_1} z_{k_2}\right]| \le$$

$$\le \alpha^2 c^2 \left(\dim\psi_k\right)^2 \frac{\psi_{\max}^2}{N}\left[|r_z(0)| + 2\sum_{\tau=1}^{N}\left(1-\frac{\tau}{N}\right)|r_z(\tau)|\right] \le$$

$$\le \frac{C}{N}\sum_{\tau=0}^{\infty} |r_z(\tau)|,$$

where

$$r_z(\tau) = var\varepsilon \sum_{i=0}^{\infty} \omega_i \omega_{i+\tau},$$

$$C = 2\alpha^2 c^2 \left(\dim\psi_k\right)^2 \psi_{\max}^2.$$

Since

$$\left|var\varepsilon \sum_{\tau=0}^{\infty}\sum_{i=0}^{\infty} \omega_i\omega_{i+\tau}\right| \le var\varepsilon \sum_{\tau=0}^{\infty}\sum_{i=0}^{\infty} |\omega_i|\,|\omega_{i+\tau}| \le var\varepsilon \sum_{i=0}^{\infty}|\omega_i| \sum_{i=0}^{\infty}|\omega_{i+\tau}| < \infty,$$

we conclude that

$$E\xi_N^2 \le D\frac{1}{N}$$

where $D = Cvar\varepsilon \left|\sum_{\tau=0}^{\infty}\sum_{i=0}^{\infty} \omega_i\omega_{i+\tau}\right|$. ∎

A.21   Proof of Theorem 5.4

Proof. To simplify presentation let $z_{\max} = 1$. From (5.50) we get

$$\left\|\Delta_N^{(IV)}(\Psi_N)\right\|^2 = \Delta_N^{(IV)T}(\Psi_N)\Delta_N^{(IV)}(\Psi_N) = Z_N^{*T}\Gamma_N^T\Gamma_N Z_N^*,$$

and the maximum value of cumulated error is

$$Q(\Psi_N) = \max_{\|\mathbf{Z}_N^*\| \leq 1} \left(\Delta_N^{(IV)T}(\Psi_N) \Delta_N^{(IV)}(\Psi_N) \right)$$

$$= \max_{\|Z_N^*\| \leq 1} \left\langle Z_N^*, \Gamma_N^T \Gamma_N Z_N^* \right\rangle$$

$$= \|\Gamma_N\|^2 = \lambda_{\max} \left(\Gamma_N^T \Gamma_N \right),$$

where $\| \ \|$ is the spectral matrix norm induced by the Euclidean vector norm, and $\lambda_{\max}()$ denotes the biggest eigenvalue of the matrix. Since [112],[154]

$$\lambda_{\max} \left(\Gamma_N^T \Gamma_N \right) = \lambda_{\max} \left(\Gamma_N \Gamma_N^T \right),$$

from definition of Γ_N (page 127) we conclude that

$$\max_{\|\mathbf{Z}_N^*\| \leq 1} \left(\Delta_N^{(IV)T}(\Psi_N) \Delta_N^{(IV)}(\Psi_N) \right) = \max_{\|\zeta\| \leq 1} \left\langle \zeta, \Gamma_N \Gamma_N^T \zeta \right\rangle =$$

$$= \max_{\|\zeta\| \leq 1} \left\langle \zeta, \left(\frac{1}{N} \Psi_N^T \Phi_N \right)^{-1} \left(\frac{1}{N} \Psi_N^T \Psi_N \right) \left(\frac{1}{N} \Phi_N^T \Psi_N \right)^{-1} \zeta \right\rangle.$$

On the basis of (5.49), it holds that

$$\max_{\|\mathbf{Z}_N^*\| \leq 1} \left(\Delta_N^{(IV)T}(\Psi_N) \Delta_N^{(IV)}(\Psi_N) \right) =$$

$$= \max_{\|\zeta\| \leq 1} \left\langle \zeta, \left(\frac{1}{N} \Psi_N^T \Phi_N^\# \right)^{-1} \left(\frac{1}{N} \Psi_N^T \Psi_N \right) \left(\frac{1}{N} \Phi_N^{\#T} \Psi_N \right)^{-1} \zeta \right\rangle,$$

with probability 1, as $N \to \infty$, where Φ_N and $\Phi_N^\#$ are given by (5.9) and (5.48), respectively. Using Lemma B.3 (page 223) for $M_1 = \frac{1}{\sqrt{N}} \Phi_N^\#$ and $M_2 = \frac{1}{\sqrt{N}} \Psi_N$ we get

$$\zeta^T \Gamma_N \Gamma_N^T \zeta \geq \zeta^T \left(\frac{1}{N} \Phi_N^{\#T} \Phi_N^\# \right)^{-1} \zeta$$

for each vector ζ, and consequently

$$Q(\Psi_N) = \max_{\|\zeta\| \leq 1} \left(\zeta^T \Gamma_N \Gamma_N^T \zeta \right) \geq \max_{\|\zeta\| \leq 1} \left(\zeta^T \left(\frac{1}{N} \Phi_N^{\#T} \Phi_N^\# \right)^{-1} \zeta \right).$$

For $\Psi_N = \Phi_N^\#$, it holds that

$$\max_{\|\zeta\| \leq 1} \left(\zeta^T \Gamma_N \Gamma_N^T \zeta \right) = \max_{\|\zeta\| \leq 1} \left(\zeta^T \left(\frac{1}{N} \Phi_N^{\#T} \Phi_N^\# \right)^{-1} \zeta \right)$$

and the criterion $Q(\Psi_N)$ attains minimum. The choice $\Psi_N = \Phi_N^\#$ is thus asymptotically optimal.   ∎

A.22   Proof of Theorem 5.7

Proof. The estimation error (5.64) can be decomposed as follows

$$\Delta_{N,M}^{(IV)} = \widehat{\theta}_{N,M}^{*(IV)} - \theta = \widehat{\theta}_{N,M}^{*(IV)} - \widehat{\theta}_N^{*(IV)} + \widehat{\theta}_N^{*(IV)} - \theta,$$

where $\widehat{\theta}_N^{*(IV)} = \left(\Psi_N^{*T}\Phi_N\right)^{-1}\Psi_N^{*T}Y_N$, and Ψ_N^* is defined by (5.52) and (5.48). From the triangle inequality, for each norm $\|\ \|$ it holds that

$$\left\|\Delta_{N,M}^{(IV)}\right\| \le \left\|\widehat{\theta}_{N,M}^{*(IV)} - \widehat{\theta}_N^{*(IV)}\right\| + \left\|\widehat{\theta}_N^{*(IV)} - \theta\right\|. \tag{A.43}$$

On the basis of Theorem 5.1 (page 120)

$$\left\|\widehat{\theta}_N^{*(IV)} - \theta\right\| \to 0 \text{ in probability}$$

as $N \to \infty$. To prove 5.7, let us analyze the component $\left\|\widehat{\theta}_{N,M}^{*(IV)} - \widehat{\theta}_N^{*(IV)}\right\|$ in (A.43) to show that, for fixed N, it tends to zero in probability, as $M \to \infty$. let us denote (cf. [63], pages 116-117)

$$\varepsilon_N \triangleq \frac{1}{\left\|\frac{1}{N}\Psi_N^{*T}\Phi_N\right\|} \quad (N-\text{fixed}).$$

From (5.62) we have that

$$\left\|\left(\frac{1}{N}\widehat{\Psi}_{N,M}^{*T}\Phi_N\right) - \left(\frac{1}{N}\Psi_N^{*T}\Phi_N\right)\right\| \to 0 \text{ in probability}$$

as $M \to \infty$, and particularly

$$\lim_{M\to\infty} P\left\{\left\|\left(\frac{1}{N}\widehat{\Psi}_{N,M}^{*T}\Phi_N\right) - \left(\frac{1}{N}\Psi_N^{*T}\Phi_N\right)\right\| < \varepsilon_N\right\} = 1.$$

Introducing

$$r_M \triangleq \frac{\left\|\left(\frac{1}{N}\widehat{\Psi}_{N,M}^{*T}\Phi_N\right) - \left(\frac{1}{N}\Psi_N^{*T}\Phi_N\right)\right\|}{\varepsilon_N\left(\varepsilon_N - \left\|\left(\frac{1}{N}\widehat{\Psi}_{N,M}^{*T}\Phi_N\right) - \left(\frac{1}{N}\Psi_N^{*T}\Phi_N\right)\right\|\right)}$$

and using Banach Theorem (see [80], Theorem 5.8., page. 106) we get

$$\lim_{M\to\infty} P\left\{\left\|\left(\frac{1}{N}\widehat{\Psi}_{N,M}^{*T}\Phi_N\right)^{-1} - \left(\frac{1}{N}\Psi_N^{*T}\Phi_N\right)^{-1}\right\| \le r_M\right\} = 1.$$

Since $r_M \to 0$ in probability as $M \to \infty$, we finally conclude that

$$\left\| \widehat{\theta}_{N,M}^{*(IV)} - \widehat{\theta}_{N}^{*(IV)} \right\| \to 0 \text{ in probability,}$$

as $M \to \infty$, for each N.   ∎

A.23   Example of ARX(1) Object

Example A.1 *Consider the simple ARX(1) linear dynamic object with the input u_k and the output v_k, disturbed by the random process ε_k, i.e.*

$$v_k = bu_k + av_{k-1}, \text{ and } y_k = v_k + \varepsilon_k.$$

Since the noise-free output v_k cannot be observed, we base on the difference equation describing dependence between u_k and the measured output y_k

$$y_k = bu_k + ay_{k-1} + z_k = (a,b)\phi_k + z_k,$$

where $\phi_k = (y_{k-1}, u_k)^T$ and the resulting disturbance

$$z_k = \varepsilon_k - a\varepsilon_{k-1}$$

is obviously correlated with y_{k-1}, included in the 'generalized input' ϕ_k.

Appendix B
Algebra Toolbox

B.1 SVD Decomposition

Theorem B.1. *[77] For each $A \in R^{m,n}$ it exists the unitary matrices $U \in R^{m,m}$, and $V \in R^{n,n}$, such that*

$$U^T A V = \Sigma = diag(\sigma_1, ..., \sigma_l), \qquad (B.1)$$

where $l = \min(m, n)$, and

$$\sigma_1 \geq \sigma_2 \geq ... \geq \sigma_r > 0$$
$$\sigma_{r+1} = ... = \sigma_l = 0$$

where $r = rank(A)$.

The numbers $\sigma_1, ..., \sigma_l$ are called the singular values of the matrix A. Solving (B.1) with respect to A we obtain

$$A = U \Sigma V^T = \sum_{i=1}^{r} u_i \sigma_i v_i^T = \sum_{i=1}^{r} \sigma_i u_i v_i^T, \qquad (B.2)$$

where u_i and v_i denote i-th columns of U and V, respectively [77].

B.2 Factorization Theorem

Theorem B.2. *[112] Each positive definite matrix M can be shown in the form*

$$M = PP^T$$

where P (root of M) is nonsingular.

Proof. Let w_i and λ_i $(i = 1, 2, ..., s)$ be the eigenvectors and eigenvalues of the matrix M, respectively. Obviously, it holds that

$$Mw_i = \lambda_i w_i.$$

Since $M = M^T$, λ_i's are real and w_i's are orthogonal. Moreover, since M is positive definite ($M > 0$), it holds that

$$\lambda_i > 0 \text{ for each } i = 1, 2, ..., s.$$

Introducing

$$W = [w_1, w_2, ..., w_s] \text{ and } \Lambda = diag(\lambda_i) = \begin{bmatrix} \lambda_1 & 0 & .. & 0 \\ 0 & \lambda_2 & 0 & .. \\ .. & 0 & .. & 0 \\ 0 & .. & 0 & \lambda_s \end{bmatrix}$$

we obtain that

$$W^T W = I, \text{ i.e., } W^{-1} = W^T.$$

Hence, $MW = W\Lambda$, and thus

$$M = W\Lambda W^{-1} = W\Lambda W^T.$$

By decomposing Λ in the following way

$$\Lambda = \Lambda^{1/2}\Lambda^{1/2}, \text{ where } \Lambda^{1/2} = diag(\sqrt{\lambda_i}) = \begin{bmatrix} \sqrt{\lambda_1} & 0 & & .. & 0 \\ 0 & \sqrt{\lambda_2} & 0 & .. \\ .. & 0 & & .. & 0 \\ 0 & & .. & 0 & \sqrt{\lambda_s} \end{bmatrix},$$

we get

$$W\Lambda W^T = W\Lambda^{1/2} \left(W\Lambda^{1/2} \right)^T$$

and

$$M = PP^T, \text{ where } P = W\Lambda^{1/2}.$$

∎

B.3 Technical Lemmas

Lemma B.1. *If $\alpha_N = O(a_N)$ in probability and $\beta_N = O(b_N)$ in probability then $\alpha_N \beta_N = O(\max\{a_N^2, b_N^2\})$ in probability and $\alpha_N + \beta_N = O(\max\{a_N, b_N\})$ in probability, where $\{a_N\}$, $\{b_N\}$ are positive number sequences convergent to zero.*

Proof. The inequality $|a|\,|b| \leqslant \frac{1}{2}(a^2 + b^2)$ implies $|\alpha_N \beta_N| / \max\{a_N^2, b_N^2\} \leqslant (1/2)\left(\alpha_N^2/a_N^2 + \beta_N^2/b_N^2\right)$ which yields

$$P\{|d_N| \frac{|\alpha_N \beta_N|}{\max\{a_N^2, b_N^2\}} > \varepsilon\}$$

$$\leqslant P\{|d_N|^{\frac{1}{2}} \frac{|\alpha_N|}{a_N} > \sqrt{\varepsilon}\} + P\{|d_N|^{\frac{1}{2}} \frac{|\beta_N|}{b_N} > \sqrt{\varepsilon}\}$$

for each $\varepsilon > 0$. Similarly $|\alpha_N + \beta_N| / \max\{a_N, b_N\} \leqslant |\alpha_N| / a_N + |\beta_N| / b_N$ and hence

$$P\{|d_N| \frac{|\alpha_N + \beta_N|}{\max\{a_N, b_N\}} > \varepsilon\}$$

$$\leqslant P\{|d_N| \frac{|\alpha_N|}{a_N} > \frac{\varepsilon}{2}\} + P\{|d_N| \frac{|\beta_N|}{b_N} > \frac{\varepsilon}{2}\}$$

each $\varepsilon > 0$, which ends the proof. ∎

Lemma B.2. *[69] If $\alpha_N = \alpha + O(a_N)$ in probability, $\beta_N = \beta + O(b_N)$ in probability and $\beta \neq 0$ then $\alpha_N/\beta_N = \alpha/\beta + O(\max\{a_N, b_N\})$ in probability.*

Lemma B.3. *Let M_1 and M_2 be two matrices with the same dimensions. If $(M_1^T M_1)^{-1}$, $(M_1^T M_2)^{-1}$ and $(M_2^T M_1)^{-1}$ exist, then*

$$D_N = (M_2^T M_1)^{-1} M_2^T M_2 (M_1^T M_2)^{-1} - (M_1^T M_1)^{-1}$$

is nonnegative definite, i.e., for each ζ it holds that

$$\zeta^T D_N \zeta \geq 0.$$

Proof. For the proof see [154]. ∎

Lemma B.4. *If λ_i, $i = 1, 2, ..., m$, are eigenvalues of the matrix A then the matrix $A - \delta I$ has the eigenvalues $\lambda_i - \delta$, $i = 1, 2, ..., m$.*

Proof. The conclusion follows immediately from the relation

$$\det([A - \delta I] - \lambda I) = \det(A - (\lambda + \delta) I)$$

∎

B.4   Types of Convergence of Random Sequences

Definition B.1. *[24] The sequence of random variables $\{\varkappa_k\}$ converges, for $k \to \infty$, with probability 1 (strongly) to $\varkappa^*$, if*

$$P(\lim_{k \to \infty} \varkappa_k = \varkappa^*) = 1.$$

Definition B.2. *[24] The sequence of random variables $\{\varkappa_k\}$ converges, for $k \to \infty$, in probability (weakly) to $\varkappa^\#$, if*

$$\lim_{k\to\infty} P(\left| \varkappa_k - \varkappa^\# \right| > \varepsilon) = 0,$$

for each $\varepsilon > 0$. The $\varkappa^\#$ is denoted as a probabilistic limit of $\varkappa_k$

$$Plim_{k\to\infty}\varkappa_k = \varkappa^\#. \tag{B.3}$$

Lemma B.5. If $\varkappa_k \to \varkappa$ with probability 1, as $k \to \infty$, then $\varkappa_k \to \varkappa$ in probability, as $k \to \infty$.

Definition B.3. [24] The sequence of random variables $\{\varkappa_k\}$ converges, for $k \to \infty$, in the mean square sense to $\varkappa^*$, if

$$\lim_{k\to\infty} E(\varkappa_k - \varkappa^*)^2 = 0.$$

Definition B.4. [47] The sequence of random variables $\{\varkappa_k\}$ has the rate of convergence $O(e_k)$ in probability as $k \to \infty$, where $\{e_k\}$ is deterministic number sequence which tends to zero, i.e.,

$$\varkappa_k = O(e_k) \text{ in probability,}$$

if $\left\{ \frac{\varkappa_k}{e_k}\chi_k \right\} \to 0$ in probability for each number sequence $\{\chi_k\}$, such that $\lim_{k\to\infty} \chi_k = 0$.

Definition B.5. [47] The sequence of random variables $\{\varkappa_k\}$ has the rate of convergence $O(e_k)$ in the mean square sense, as $k \to \infty$, if it exists the constant $0 \le c < \infty$, such that

$$E\varkappa_k^2 \le ce_k.$$

Lemma B.6. [47] If $\varkappa_k = O(e_k)$ in the mean square sense, then $\varkappa_k = O(\sqrt{e_k})$ in probability.

B.5 Slutzky Theorem

Theorem B.3. ([112]) If $Plim_{k\to\infty}\varkappa_k = \varkappa^\#$ and the function $g()$ is continuous, then

$$P\lim_{k\to\infty} g(\varkappa_k) = g(\varkappa^\#).$$

B.6 Chebychev's Inequality

Lemma B.7. ([24], page 106) For each constant c, each random variable X and each $\varepsilon > 0$ it holds that

$$P\{|X - c| > \varepsilon\} \le \frac{1}{\varepsilon^2}E(X - c)^2.$$

In particular, for $c = EX$

$$P\{|X - EX| > \varepsilon\} \leq \frac{1}{\varepsilon^2} var X.$$

B.7 Persistent Excitation

Definition B.6. *([131]) The stationary random process $\{\alpha_k\}$ is strongly persistently exciting of orders $n \times m$, (denote $SPE(n, m)$) if the matrix*

$$R_{\varkappa}(n, m) = E \begin{bmatrix} \varkappa_k \\ \vdots \\ \varkappa_{k-n+1} \end{bmatrix} \begin{bmatrix} \varkappa_k \\ \vdots \\ \varkappa_{k-n+1} \end{bmatrix}^T$$

where $\varkappa_k = \begin{bmatrix} \alpha_k & \alpha_k^2 & .. & \alpha_k^m \end{bmatrix}^T$, is of full rank.

Lemma B.8. *([131]) The i.i.d. process $\{\alpha_k\}$ is $SPE(n, m)$ for each n and m.*

Lemma B.9. *([131]) Let $x_k = H(q^{-1})u_k$, $H(q^{-1})$ be asymptotically stable linear filter, and $\{u_k\}$ be a random sequence with finite variance. If the frequency function of $\{u_k\}$ is strictly positive in at least $m + 1$ distinct points, then $\{x_k\}$ is $SPE(n, m)$ for each n.*

B.8 Ergodic Processes

Definition B.7. *([128]) The stationary stochastic process $\{\varkappa_k\}$ is ergodic with respect to the first and the second order moments if*

$$\frac{1}{N} \sum_{k=1}^{N} \varkappa_k \to E\varkappa_k$$

$$\frac{1}{N} \sum_{k=1}^{N} \varkappa_k \varkappa_{k+\tau} \to E\varkappa_k \varkappa_{k+\tau}$$

with probability 1, as $N \to \infty$.

Lemma B.10. *(see [105], or [128]) Let us assume that $\{\varkappa_k\}$ is a discrete-time random process with finite variance. If the autocorrelation of $\{\varkappa_k\}$ is such that $r_{\varkappa}(\tau) \to 0$ for $|\tau| \to \infty$, then*

$$\frac{1}{N} \sum_{k=1}^{N} \varkappa_k \to E\varkappa \qquad (B.4)$$

with probability 1, as $N \to \infty$.

Lemma B.11. *(cf. [105], or [128]) If the two random processes $\{\varkappa_{1,k}\}$ and $\{\varkappa_{2,k}\}$ have finite fourth order moments and $r_{\varkappa_1}(\tau) \to 0$, $r_{\varkappa_2}(\tau) \to 0$ as $|\tau| \to \infty$, then*

$$\frac{1}{N} \sum_{k=1}^{N} \varkappa_{1,k} \varkappa_{2,k} \to E \varkappa_{1,k} \varkappa_{2,k}$$

with probability 1, as $N \longrightarrow \infty$.

B.9 Modified Triangle Inequality

Lemma B.12. *[24] If X and Y are k-dimensional random vectors, then $P\left[\|X + Y\| \geqslant \varepsilon\right] \leqslant P\left[\|X\| \geqslant \frac{\varepsilon}{2}\right] + P\left[\|Y\| \geqslant \frac{\varepsilon}{2}\right]$ for each vector norm $\|\bullet\|$ and each $\varepsilon > 0$.*

Proof. Let us define the following random events: A: $\|X + Y\| \geqslant \varepsilon$, B: $\|X\| + \|Y\| \geqslant \varepsilon$, C: $\|X\| \geqslant \frac{\varepsilon}{2}$, D: $\|Y\| \geqslant \frac{\varepsilon}{2}$. Obviously $A \Longrightarrow B$ and $B \Longrightarrow (C \smile D)$. Thus $A \subset B \subset (C \smile D)$ and $P(A) \leqslant P(B) \leqslant P(C \smile D) \leqslant P(C) + P(D)$. ∎

References

[1] Akaike, H.: A new look at the statistical model identification. IEEE Transactions on Automatic Control 19(6), 716–723 (1974) (*cit.* pp. 149 and 191)

[2] Bai, E.W.: An optimal two-stage identification algorithm for Hammerstein-Wiener nonlinear systems. Automatica 34(3), 333–338 (1998) (*cit.* pp. 24, 32, 113, 115, 116, 117, and 118)

[3] Bai, E.W.: A blind approach to the Hammerstein–Wiener model identification. Automatica 38(6), 967–979 (2002) (*cit.* pp. 14 and 87)

[4] Bai, E.W.: Frequency domain identification of Hammerstein models. IEEE Transactions on Automatic Control 48(4), 530–542 (2003) (*cit.* pp. 26 and 27)

[5] Bai, E.W.: Frequency domain identification of Wiener models. Automatica 39(9), 1521–1530 (2003) (*cit.* pp. 26, 27, and 87)

[6] Bai, E.W., Fu, M.: A blind approach to Hammerstein model identification. IEEE Transactions on Signal Processing 50(7), 1610–1619 (2002) (*cit.* p. 23)

[7] Bai, E.W., Li, D.: Convergence of the iterative Hammerstein system identification algorithm. IEEE Transactions on Automatic Control 49(11), 1929–1940 (2004) (*cit.* p. 14)

[8] Bai, E.W., Reyland Jr., J.: Towards identification of Wiener systems with the least amount of a priori information: IIR cases. Automatica 45(4), 956–964 (2009) (*cit.* p. 87)

[9] Bershad, N.J., Bouchired, S., Castanie, F.: Stochastic analysis of adaptive gradient identification of Wiener-Hammerstein systems for Gaussian inputs. IEEE Transactions on Signal Processing 48(2), 557–560 (2000) (*cit.* p. 87)

[10] Bershad, N.J., Celka, P., Vesin, J.M.: Analysis of stochastic gradient tracking of time-varying polynomial Wiener systems. IEEE Transactions on Signal Processing 48(6), 1676–1686 (2000) (*cit.* p. 87)

[11] Billings, S.A.: Structure detection and model validity tests in the identification of nonlinear systems. IEEE Proceedings 130(4), 193–200 (1983) (*cit.* pp. 149 and 191)

[12] Billings, S.A., Fakhouri, S.Y.: Identification of nonlinear systems using the Wiener model. Electronics Letters 13(17), 502–504 (1977) (*cit.* pp. 15 and 16)

[13] Billings, S.A., Fakhouri, S.Y.: Identification of systems containing linear dynamic and static nonlinear elements. Automatica 18(1), 15–26 (1982) (*cit.* pp. 14, 15, 24, and 34)

[14] Bodson, M., Chiasson, J.N., Novotnak, R.T., Rekowski, R.B.: High-performance nonlinear feedback control of a permanent magnet stepper motor. IEEE Transactions on Control Systems Technology 1(1), 5–14 (1993) (*cit.* p. 121)

[15] Box, G.E.P., Jenkins, G.M., Reinsel, G.C.: Time Series Analysis. Holden-day San Francisco (1976) (*cit.* pp. 79 and 209)

[16] Brown, R.G., Hwang, P.: Introduction to Random Signals and Applied Kalman Filtering. John Wiley & Sons, New York (1992) (*cit.* p. 171)

[17] Cadzow, J.A.: Blind deconvolution via cumulant extrema. IEEE Signal Processing Magazine 13(3), 24–42 (1996) (*cit.* pp. 64, 67, 77, and 79)

[18] Celka, P., Bershad, N.J., Vesin, J.M.: Stochastic gradient identification of polynomial Wiener systems: Analysis and application. IEEE Transactions on Signal Processing 49(2), 301–313 (2001) (*cit.* p. 103)

[19] Chang, F., Luus, R.: A noniterative method for identification using Hammerstein model. IEEE Transactions on Automatic Control 16(5), 464–468 (1971) (*cit.* pp. 14 and 15)

[20] Chen, B.M., Lee, T.H., Peng, K., Venkataramanan, V.: Composite nonlinear feedback control for linear systems with input saturation: theory and an application. IEEE Transactions on Automatic Control 48(3), 427–439 (2003) (*cit.* p. 121)

[21] Chen, S., Billings, S.A.: Representations of non-linear systems: the NARMAX model. International Journal of Control 49(3), 1013–1032 (1989) (*cit.* p. 113)

[22] Chen, S., Billings, S.A., Luo, W.: Orthogonal least squares methods and their application to non-linear system identification. International Journal of Control 50(5), 1873–1896 (1989) (*cit.* p. 171)

[23] Cheney, E.W.: Introduction to Approximation Theory. AMS Chelsea Publishing, Rhode Island (1982) (*cit.* pp. 24 and 28)

[24] Chow, Y.S., Teicher, H.: Probability Theory: Independence, Interchangeability, Martingales. Springer (2003) (*cit.* pp. 215, 223, 224, and 226)

[25] Cristobal, J.A., Roca, P.F., Manteiga, W.G.: A class of linear regression parameter estimators constructed by nonparametric estimation. The Annals of Statistics 15(2), 603–609 (1987) (*cit.* pp. 27 and 29)

[26] Darling, R.B.: A full dynamic model for pn-junction diode switching transients. IEEE Transactions on Electron Devices 42(5), 969–976 (1995) (*cit.* p. 162)

[27] Devore, R.A.: Constructive Approximation. Springer, Heidelberg (1993) (*cit.* p. 150)

[28] Devore, R.A.: Nonlinear approximation. Acta Numerica, 51–150 (1998) (*cit.* p. 150)

[29] Ding, F., Chen, T.: Identification of Hammerstein nonlinear ARMAX systems. Automatica 41(9), 1479–1489 (2005) (*cit.* p. 33)

[30] Efromovich, S.: Nonparametric Curve Estimation. Springer, New York (1999) (*cit.* pp. 27 and 49)

[31] Feng, C.C., Chi, C.Y.: Performance of cumulant based inverse filters for blind deconvolution. IEEE Transactions on Signal Processing 47(7), 1922–1935 (1999) (*cit.* pp. 64 and 77)

[32] Findeisen, W., Bailey, F.N., Brdyś, M., Malinowski, K., Tatjewski, P., Woźniak, A.: Control and Coordination in Hierarchical Systems. J. Wiley, Chichester (1980) (*cit.* pp. 137, 147, and 184)

[33] Finigan, B., Rowe, I.: Strongly consistent parameter estimation by the introduction of strong instrumental variables. IEEE Transactions on Automatic Control 19(6), 825–830 (1974) (*cit.* pp. 56 and 119)

[34] Fitzgerald, W.J., Smith, R.L., Walden, A.T., Young, P.C.: Nonlinear and Nonstationary Signal Processing. Cambridge University Press (2001) (*cit.* p. 171)

[35] Giannakis, G.B., Serpedin, E.: A bibliography on nonlinear system identification. Signal Processing 81(3), 533–580 (2001) (*cit.* pp. 21 and 103)

[36] Giri, F., Bai, E.W.: Block-Oriented Nonlinear System Identification. LNCIS, vol. 404. Springer, Heidelberg (2010) (*cit.* pp. 14 and 113)

[37] Giri, F., Rochdi, Y., Chaoui, F.Z.: An analytic geometry approach to Wiener system frequency identification. IEEE Transactions on Automatic Control 54(4), 683–696 (2009) (*cit.* p. 87)

[38] Giri, F., Rochdi, Y., Chaoui, F.Z.: Parameter identification of Hammerstein systems containing backlash operators with arbitrary-shape parametric borders. Automatica 47(8), 1827–1833 (2011) (*cit.* p. 23)

[39] Gonçalves, J.M., Megretski, A., Dahleh, M.A.: Global stability of relay feedback systems. IEEE Transactions on Automatic Control 46(4), 550–562 (2001) (*cit.* p. 121)

[40] Greblicki, W.: Nonparametric orthogonal series identification of Hammerstein systems. International Journal of System Science 20, 2355–2367 (1989) (*cit.* pp. 13, 16, 23, 27, 28, 49, and 78)

[41] Greblicki, W.: Nonparametric identification of Wiener systems. IEEE Transactions on Information Theory 38(5), 1487–1493 (1992) (*cit.* p. 87)

[42] Greblicki, W.: Nonparametric approach to Wiener system identification. IEEE Transactions on Circuits and Systems I: Fundamental Theory and Applications 44(6), 538–545 (1997) (*cit.* pp. 87, 88, and 89)

[43] Greblicki, W.: Nonlinearity recovering in Wiener system driven with correlated signal. IEEE Transactions on Automatic Control 49(10), 1805–1812 (2004) (*cit.* p. 89)

[44] Greblicki, W.: Nonparametric input density-free estimation of the nonlinearity in Wiener systems. IEEE Transactions on Information Theory 56(7), 3575–3580 (2010) (*cit.* pp. 18, 87, 103, 107, and 108)

[45] Greblicki, W., Krzyżak, A., Pawlak, M.: Distribution-free pointwise consistency of kernel regression estimate. The Annals of Statistics 12(4), 1570–1575 (1984) (*cit.* pp. 11 and 178)

[46] Greblicki, W., Mzyk, G.: Semiparametric approach to Hammerstein system identification. In: Proceedings of the 15th IFAC Symposium on System Identification, Saint-Malo, France, pp. 1680–1685 (2009) (*cit.* pp. 56, 82, 141, and 169)

[47] Greblicki, W., Pawlak, M.: Fourier and Hermite series estimates of regression functions. Annals of the Institute of Statistical Mathematics 37(1), 443–454 (1985) (*cit.* p. 224)

[48] Greblicki, W., Pawlak, M.: Identification of discrete Hammerstein systems using kernel regression estimates. IEEE Transactions on Automatic Control 31, 74–77 (1986) (*cit.* pp. 14, 16, 23, 27, 28, 30, 31, 33, 42, 48, 49, 50, 80, 84, and 209)

[49] Greblicki, W., Pawlak, M.: Cascade non-linear system identification by a non-parametric method. International Journal of System Science 25, 129–153 (1994) (*cit.* pp. 14, 16, 27, 28, 30, 31, 33, 49, 50, 51, 55, 80, and 209)

[50] Greblicki, W., Pawlak, M.: Nonparametric System Identification. Cambridge University Press (2008) (*cit.* pp. VII, 10, 11, 12, 13, 16, 27, 30, 57, 71, 75, 78, 89, 90, 91, 100, 103, 111, 129, 156, and 191)

[51] Greblicki, W., Rutkowska, D., Rutkowski, L.: An orthogonal series estimate of time-varying regression. Annals of the Institute of Statistical Mathematics 35(1), 215–228 (1983) (*cit.* pp. 171 and 179)

[52] Györfi, L., Kohler, M., Krzyżak, A., Walk, H.: A Distribution-Free Theory of Nonparametric Regression. Springer, New York (2002) (*cit.* p. 11)

[53] Haber, R.: Structural identification of quadratic block-oriented models based on estimated Volterra kernels. International Journal of Systems Science 20(8), 1355–1380 (1989) (*cit.* p. 14)

[54] Haber, R., Keviczky, L.: Nonlinear System Identification: Input-Output Modeling Approach. Kluwer Academic Publishers (1999) (*cit.* pp. 16, 23, 24, 113, 149, and 191)

[55] Hagenblad, A., Ljung, L., Wills, A.: Maximum likelihood identification of Wiener models. Automatica 44(11), 2697–2705 (2008) (*cit.* p. 87)

[56] Haist, N.D., Chang, F.H., Luus, R.: Nonlinear identification in the presence of correlated noise using Hammerstein model. IEEE Transactions on Automatic Control 18, 552–555 (1973) (*cit.* pp. 24 and 28)

[57] Hannan, E.J., Deistler, M.: The Statistical Theory of Linear Systems. John Wiley and Sons (1988) (*cit.* pp. 37 and 197)

[58] Hansen, L.P., Singleton, K.J.: Generalized instrumental variables estimation of nonlinear rational expectations models. Econometrica: Journal of the Econometric Society, 1269–1286 (1982) (*cit.* p. 119)

[59] Härdle, W.: Applied Nonparametric Regression. Cambridge University Press, Cambridge (1990) (*cit.* pp. 27, 28, and 49)

[60] Härdle, W., Kerkyacharian, G., Picard, D., Tsybakov, A.: Wavelets, Approximation, and Statistical Applications. Springer, New York (1998) (*cit.* p. 27)

[61] Hasiewicz, Z.: Identification of a linear system observed through zero-memory non-linearity. International Journal of Systems Science 18(9), 1595–1607 (1987) (*cit.* pp. 16, 137, and 138)

[62] Hasiewicz, Z.: Applicability of least-squares to the parameter estimation of large-scale no-memory linear composite systems. International Journal of Systems Science 20(12), 2427–2449 (1989) (*cit.* pp. 56, 57, 137, 138, 139, and 141)

[63] Hasiewicz, Z.: Identyfikacja Sterowanych Systemów o Złożonej Strukturze. Wydawn. Politechniki Wrocławskiej (1993) (in Polish) (*cit.* p. 219)

[64] Hasiewicz, Z., Mzyk, G.: Combined parametric-nonparametric identification of Hammerstein systems. IEEE Transactions on Automatic Control 48(8), 1370–1376 (2004) (*cit.* pp. 23, 35, 45, 53, 55, 67, 71, 77, 91, 96, and 152)

[65] Hasiewicz, Z., Mzyk, G.: Hammerstein system identification by nonparametric instrumental variables. International Journal of Control 82(3), 440–455 (2009) (*cit.* pp. 33, 56, 57, 75, 77, 79, 84, 91, 96, 98, 111, 119, 143, 152, 161, and 207)

[66] Hasiewicz, Z., Mzyk, G., Śliwiński, P.: Quasi-parametric recovery of Hammerstein system nonlinearity by smart model selection. In: Rutkowski, L., Scherer, R., Tadeusiewicz, R., Zadeh, L.A., Zurada, J.M. (eds.) ICAISC 2010, Part II. LNCS (LNAI), vol. 6114, pp. 34–41. Springer, Heidelberg (2010) (*cit.* p. 33)

[67] Hasiewicz, Z., Mzyk, G., Śliwiński, P., Wachel, P.: Mixed parametric-nonparametric identification of Hammerstein and Wiener systems-a survey. In: Proceedings of the 16th IFAC Symposium on System Identification, Brussels, Belgium, vol. 16, pp. 464–469 (2012) (*cit.* pp. 23 and 96)

[68] Hasiewicz, Z., Mzyk, G., Śliwiński, P., Wachel, P.: Model selection of Hammerstein system nonlinearity under heavy noise. In: Giri, F., Van Assche, V. (eds.) 11th IFAC International Workshop on Adaptation and Learning in Control and Signal Processing, Caen, France, July 3-5, pp. 378–383. International Federation of Automatic Control (2013) (*cit.* p. 149)

[69] Hasiewicz, Z., Pawlak, M., Śliwiński, P.: Nonparametric identification of nonlinearities in block-oriented systems by orthogonal wavelets with compact support. IEEE Transactions on Circuits and Systems I: Regular Papers 52(2), 427–442 (2005) (*cit.* pp. 13, 27, 28, 49, 59, 80, and 223)

[70] Hasiewicz, Z., Śliwiński, P.: Identification of non-linear characteristics of a class of block-oriented non-linear systems via Daubechies wavelet-based models. International Journal of System Science 33(14), 1121–1144 (2002) (*cit.* pp. 23, 27, and 28)

[71] Hasiewicz, Z., Śliwiński, P., Mzyk, G.: Nonlinear system identification under various prior knowledge. In: Proceedings of the 17th World Congress. The International Federation of Automatic Control, pp. 7849–7858 (2008) (*cit.* pp. 56 and 87)

[72] Hill, D.J., Chong, C.N.: Lyapunov functions of Lur'e-Postnikov form for structure preserving models of power systems. Automatica 25(3), 453–460 (1989) (*cit.* p. 132)

[73] Hill, D.J., Mareels, I.M.Y.: Stability theory for differential/algebraic systems with application to power systems. IEEE Transactions on Circuits and Systems 37(11), 1416–1423 (1990) (*cit.* p. 132)

[74] Hunter, I.W., Korenberg, M.J.: The identification of nonlinear biological systems: Wiener and Hammerstein cascade models. Biological Cybernetics 55(2), 135–144 (1986) (*cit.* p. 103)

[75] Jeng, J.C., Huang, H.P.: Nonparametric identification for control of MIMO Hammerstein systems. Industrial & Engineering Chemistry Research 47(17), 6640–6647 (2008) (*cit.* p. 77)

[76] Keesman, K.J.: System Identification: an Introduction. Advanced Textbooks in Control and Signal Processing. Springer (2011) (*cit.* p. 1)

[77] Kiełbasiński, A., Schwetlick, H.: Numeryczna Algebra Liniowa: Wprowadzenie do Obliczeń Zautomatyzowanych. Wydawnictwa Naukowo-Techniczne, Warszawa (1992) (in Polish) (*cit.* p. 221)

[78] Kincaid, D.R., Cheney, E.W.: Numerical Analysis: Mathematics of Scientific Computing, vol. 2. Amer Mathematical Society (2002) (*cit.* p. 117)

[79] Kowalczuk, Z., Kozłowski, J.: Continuous-time approaches to identification of continuous-time systems. Automatica 36(8), 1229–1236 (2000) (*cit.* p. 119)

[80] Kudrewicz, J.: Analiza Funkcjonalna dla Automatyków i Elektroników. Państwowe Wydawnictwo Naukowe, Warszawa (1976) (in Polish) (*cit.* pp. 125 and 219)

[81] Lacy, S.L., Bernstein, D.S.: Identification of FIR Wiener systems with unknown, non-invertible, polynomial non-linearities. International Journal of Control 76(15), 1500–1507 (2003) (*cit.* pp. 18 and 87)

[82] Ljung, L.: System Identification. Wiley Online Library (1999) (*cit.* p. 171)

[83] Ljung, L., Gunnarsson, S.: Adaptation and tracking in system identification - a survey. Automatica 26(1), 7–21 (1990) (*cit.* p. 171)

[84] Ljung, L., Söderström, T.: Theory and Practice of Recursive Identification. MIT Press, Cambridge (1983) (*cit.* pp. 171, 173, and 176)

[85] Lu, J.G., Hill, D.J.: Impulsive synchronization of chaotic Lur'e systems by linear static measurement feedback: an LMI approach. IEEE Transactions on Circuits and Systems II: Express Briefs 54(8), 710–714 (2007) (*cit.* pp. 113 and 132)

[86] Moré, J.: The Levenberg-Marquardt algorithm: Implementation and theory. Lecture Notes in Mathematics, pp. 105–116 (1977) (*cit.* p. 50)

[87] Mzyk, G.: Application of instrumental variable method to the identification of Hammerstein-Wiener systems. In: Proceedings of the 6th International Conference MMAR, pp. 951–956 (2000) (*cit.* p. 113)

[88] Mzyk, G.: Kernel-based instrumental variables for NARMAX system identification. In: Proceedings of the International Conference ICSES, pp. 469–475 (2001) (*cit.* p. 113)

[89] Mzyk, G.: Zastosowanie metody zmiennych instrumentalnych do identyfikacji systemów Hammersteina-Wienera. Pomiary Automatyka Kontrola 47(7/8), 35–40 (2001) (in Polish) (*cit.* p. 113)

[90] Mzyk, G.: A censored sample mean approach to nonparametric identification of nonlinearities in Wiener systems. IEEE Transactions on Circuits and Systems – II: Express Briefs 54(10), 897–901 (2007) (*cit.* pp. 18, 87, 90, 91, 95, 96, 103, 107, 210, and 213)

[91] Mzyk, G.: Generalized kernel regression estimate for Hammerstein system identification. International Journal of Applied Mathematics and Computer Science 17(2), 101–109 (2007) (*cit.* pp. 48 and 131)

[92] Mzyk, G.: Nonlinearity recovering in Hammerstein system from short measurement sequence. IEEE Signal Processing Letters 16(9), 762–765 (2009) (*cit.* pp. 48, 66, 77, 91, 96, 131, 141, and 191)

[93] Mzyk, G.: Parametric versus nonparametric approach to Wiener systems identification. In: Giri, F., Bai, E.-W. (eds.) Block-oriented Nonlinear System Identification. LNCIS, vol. 404, pp. 111–125. Springer, Heidelberg (2010) (*cit.* pp. 90 and 103)

[94] Mzyk, G.: Wiener-Hammerstein system identification with non-Gaussian input. In: IFAC International Workshop on Adaptation and Learning in Control and Signal Processing, Antalya, Turkey (2010) (*cit.* p. 87)

[95] Mzyk, G.: Fast model order selection of the nonlinearity in Hammerstein systems. In: Proceedings of the 13th International Carpathian Control Conference (ICCC), Slovakia, pp. 507–510. IEEE (2012) (*cit.* p. 164)

[96] Mzyk, G.: Identification of interconnected systems by instrumental variables method. In: Proceedings of the 7th International Conference on Electrical and Control Technologies (ECT), Kaunas, Lithuania, pp. 13–16. IEEE (2012) (*cit.* p. 141)

[97] G. Mzyk. Nonparametric recovering of nonlinearity in Wiener-Hammerstein systems. In *Proceedings of the 9th International Conference on Informatics in Control, Automation and Robotics (ICINCO), Rome, Italy*, pages 439–445. IEEE, 2012. (*cit.* pp. 87 and 103)

[98] Mzyk, G.: Instrumental variables for nonlinearity recovering in block-oriented systems driven by correlated signal. International Journal of Systems Science (2013), doi:10.1080/00207721.2013.775682 (*cit.* pp. 56, 74, 76, 116, and 136)

[99] Mzyk, G.: Nonparametric instrumental variables for identification of block-oriented systems. International Journal of Applied Mathematics and Computer Science 23(3), 521–537 (2013) (*cit.* p. 113)

[100] Nadaraya, E.A.: On estimating regression. Teoriya Veroyatnostei i ee Primeneniya 9(1), 157–159 (1964) (*cit.* pp. VII and 16)

[101] Narendra, K.S., Gallman, P.G.: An iterative method for the identification of nonlinear systems using the Hammerstein model. IEEE Transactions on Automatic Control 11, 546–550 (1966) (*cit.* pp. 14, 15, 24, 28, and 57)

[102] Newey, W.K., Powell, J.L.: Instrumental variable estimation of nonparametric models. Econometrica 71(5), 1565–1578 (2003) (*cit.* pp. 9 and 37)

[103] Niedźwiecki, M.: First-order tracking properties of weighted least squares estimators. IEEE Transactions on Automatic Control 33(1), 94–96 (1988) (*cit.* p. 171)

[104] Niedźwiecki, M.: Identification of Time-Varying Processes. Wiley (2000) (*cit.* p. 171)

[105] Ninness, B.: Strong laws of large numbers under weak assumptions with application. IEEE Transactions on Automatic Control 45(11), 2117–2122 (2000) (*cit.* pp. 63, 225, and 226)

[106] Norton, J.P.: Optimal smoothing in the identification of linear time-varying systems. Proceedings of the Institution of Electrical Engineers 122(6), 663–668 (1975) (*cit.* p. 171)

[107] Pawlak, M., Hasiewicz, Z.: Nonlinear system identification by the Haar multiresolution analysis. IEEE Transactions on Circuits and Systems I: Fundamental Theory and Applications 45(9), 945–961 (1998)   (*cit.* pp. 14, 58, and 59)

[108] Pawlak, M., Hasiewicz, Z., Wachel, P.: On nonparametric identification of Wiener systems. IEEE Transactions on Signal Processing 55(2), 482–492 (2007) (*cit.* pp. 18, 87, 96, and 101)

[109] Pintelon, R., Schoukens, J.: System Identification: a Frequency Domain Approach. Wiley-IEEE Press (2004) (*cit.* p. 1)

[110] Press, W.H., Teukolsky, S.A., Vetterling, W.T., Flannery, B.P.: Numerical Recipes in C (1992) (*cit.* p. 50)

[111] Pupeikis, R.: On the identification of Wiener systems having saturation-like functions with positive slopes. Informatica 16(1), 131–144 (2005) (*cit.* p. 87)

[112] Rao, C.R.: Linear Statistical Inference and its Applications. Wiley, NY (1973) (*cit.* pp. 202, 218, 221, and 224)

[113] Rochdi, Y., Giri, F., Gning, J., Chaoui, F.Z.: Identification of block-oriented systems in the presence of nonparametric input nonlinearities of switch and backlash types. Automatica 46(5), 864–877 (2010) (*cit.* p. 57)

[114] Ruppert, D., Wand, M.P., Carroll, R.J.: Semiparametric Regression, vol. 12. Cambridge University Press (2003) (*cit.* p. 44)

[115] Rutkowski, L.: Nonparametric identification of quasi-stationary systems. Systems & Control letters 6(1), 33–35 (1985) (*cit.* p. 171)

[116] Rutkowski, L.: Real-time identification of time-varying systems by nonparametric algorithms based on Parzen kernels. International Journal of Systems Science 16(9), 1123–1130 (1985) (*cit.* p. 171)

[117] Rutkowski, L.: Application of multiple Fourier series to identification of multivariable non-stationary systems. International Journal of Systems Science 20(10), 1993–2002 (1989) (*cit.* p. 171)

[118] Sastry, S.: Nonlinear Systems: Analysis, Stability, and Control. Springer, New
       York (1999) (*cit.* p. 113)
[119] Schetzen, M.: The Volterra and Wiener Theories of Nonlinear Systems. John
       Wiley & Sons (1980) (*cit.* p. 14)
[120] Schoukens, J., Pintelon, R.: Identification of Linear Systems: A Practical
       Guideline to Accurate Modeling. Pergamon Press, New York (1991) (*cit.* p. 1)
[121] Schoukens, J., Pintelon, R., Dobrowiecki, T., Rolain, Y.: Identification of
       linear systems with nonlinear distortions. Automatica 41(3), 491–504 (2005)
       (*cit.* p. 15)
[122] Schoukens, J., Suykens, J., Ljung, L.: Wiener-Hammerstein benchmark. In:
       Proceedings of the 15th IFAC Symposium on System Identification (SYSID
       2009), Saint-Malo, France (2009) (*cit.* pp. 182 and 184)
[123] Seber, G.A.F., Wild, C.J.: Nonlinear Regression. John Wiley, New York
       (1989) (*cit.* p. 50)
[124] Śliwiński, P.: On-line wavelet estimation of Hammerstein system nonlinearity.
       International Journal of Applied Mathematics and Computer Science 20(3)
       (2010) (*cit.* p. 171)
[125] Śliwiński, P.: Nonlinear System Identification by Haar Wavelets. Lecture
       Notes in Statistics, vol. 210. Springer (2013) (*cit.* p. 27)
[126] Śliwiński, P., Rozenblit, J., Marcellin, M.W., Klempous, R.: Wavelet amend-
       ment of polynomial models in Hammerstein systems identification. IEEE
       Transactions on Automatic Control 54(4), 820–825 (2009) (*cit.* pp. 82, 150,
       and 169)
[127] Söderström, T., Stoica, P.: Instrumental Variable Methods for System Iden-
       tification. LNCIS, vol. 57. Springer, Heidelberg (1983) (*cit.* pp. 9, 38, 56,
       116, and 119)
[128] Söderström, T., Stoica, P.: System Identification, NJ. Prentice Hall, Engle-
       wood Cliffs (1989) (*cit.* pp. 1, 10, 26, 33, 37, 38, 194, 198, 199, 225, and 226)
[129] Söderström, T., Stoica, P.: Instrumental variable methods for system identifi-
       cation. Circuits, Systems, and Signal Processing 21(1), 1–9 (2002) (*cit.* pp. 33,
       37, 38, 56, 84, and 119)
[130] Stoica, P., Söderström, T.: Comments on the Wong and Polak minimax ap-
       proach to accuracy optimization of instrumental variable methods. IEEE
       Transactions on Automatic Control 27(5), 1138–1139 (1982) (*cit.* p. 14)
[131] Stoica, P., Söderström, T.: Instrumental-variable methods for identification of
       Hammerstein systems. International Journal of Control 35(3), 459–476 (1982)
       (*cit.* pp. 14, 38, 39, 54, 55, 57, and 225)
[132] Stone, C.J.: Optimal rates of convergence for nonparametric estimators. The
       Annals of Statistics, 1348–1360 (1980) (*cit.* p. 49)
[133] Stone, C.J.: Optimal global rates of convergence for nonparametric regression.
       The Annals of Statistics 10(4), 1040–1053 (1982) (*cit.* p. 49)
[134] Strang, G.: Introduction to Linear Algebra. Wellesley Cambridge Pr (2003)
       (*cit.* p. 47)
[135] Suykens, J., Yang, T., Chua, L.O.: Impulsive synchronization of chaotic Lur'e
       systems by measurement feedback. International Journal of Bifurcation and
       Chaos 8(06), 1371–1381 (1998) (*cit.* pp. 113 and 132)
[136] Takezawa, K.: Introduction to Nonparametric Regression. Wiley Online Li-
       brary (2006) (*cit.* p. 49)
[137] Tsatsanis, M.K., Giannakis, G.B.: Time-varying system identification and
       model validation using wavelets. IEEE Transactions on Signal Process-
       ing 41(12), 3512–3523 (1993) (*cit.* p. 171)

[138] Tugnait, J.K.: Estimation of linear parametric models using inverse filter criteria and higher order statistics. IEEE Transactions on Signal Processing 41(11), 3196–3199 (1993) (*cit.* p. 77)

[139] Van der Vaart, A.W.: Asymptotic Statistics. Cambridge University Press (2000) (*cit.* p. 206)

[140] Van Pelt, T.H., Bernstein, D.S.: Non-linear system identification using Hammerstein and non-linear feedback models with piecewise linear static maps. International Journal of Control 74(18), 1807–1823 (2001) (*cit.* p. 121)

[141] Vandersteen, G., Schoukens, J.: Measurement and identification of nonlinear systems consisting of linear dynamic blocks and one static nonlinearity. IEEE Transactions on Automatic Control 44(6), 1266–1271 (1999) (*cit.* pp. 57 and 103)

[142] Vapnik, V.N.: Estimation of Dependences Based on Empirical Data. Springer (1982) (*cit.* p. 206)

[143] Verhaegen, M., Verdult, V.: Filtering and system identification: a least squares approach. Cambridge University Press (2007) (*cit.* p. 1)

[144] Vöros, J.: Iterative algorithm for identification of Hammerstein systems with two-segment nonlinearities. IEEE Transactions on Automatic Control 44, 2145–2149 (1999) (*cit.* pp. 24 and 28)

[145] Vöros, J.: Recursive identification of Hammerstein systems with discontinuous nonlinearities containing dead-zones. IEEE Transactions on Automatic Control 48(12), 2203–2206 (2003) (*cit.* p. 57)

[146] Vörös, J.: Parameter identification of Wiener systems with multisegment piecewise-linear nonlinearities. Systems & Control Letters 56(2), 99–105 (2007) (*cit.* pp. 87 and 103)

[147] Wachel, P.: Parametric-Nonparametric Identification of Wiener Systems. PhD Thesis. Wroclaw University of Technology (2008) (in Polish) (*cit.* pp. 87, 96, 97, and 100)

[148] Wachel, P., Śliwiński, P.: Nonparametric identification of MISO Hammerstein systems with nondecomposable nonlinearities. Submitted to the IEEE Transactions on Automatic Control (2012) (*cit.* p. 191)

[149] Wand, M.P., Jones, H.C.: Kernel Smoothing. Chapman and Hall, London (1995) (*cit.* pp. 27, 30, 49, and 50)

[150] Ward, R.: Notes on the instrumental variable method. IEEE Transactions on Automatic Control 22(3), 482–484 (1977) (*cit.* pp. 33, 56, and 119)

[151] Westwick, D., Verhaegen, M.: Identifying MIMO Wiener systems using subspace model identification methods. Signal Processing 52(2), 235–258 (1996) (*cit.* pp. 87 and 103)

[152] Westwick, D.T., Schoukens, J.: Initial estimates of the linear subsystems of Wiener–Hammerstein models. Automatica (2012) (*cit.* p. 103)

[153] Wigren, T.: Convergence analysis of recursive identification algorithms based on the nonlinear Wiener model. IEEE Transactions on Automatic Control 39(11), 2191–2206 (1994) (*cit.* p. 87)

[154] Wong, K., Polak, E.: Identification of linear discrete time systems using the instrumental variable method. IEEE Transactions on Automatic Control 12(6), 707–718 (1967) (*cit.* pp. 9, 33, 56, 84, 116, 119, 127, 218, and 223)

[155] Wu, W.B., Mielniczuk, J.: Kernel density estimation for linear processes. The Annals of Statistics 30(5), 1441–1459 (2002) (*cit.* p. 11)

[156] Xiao, Y., Chen, H., Ding, F.: Identification of multi-input systems based on correlation techniques. International Journal of Systems Science 42(1), 139–147 (2011) (*cit.* p. 56)

[157] Xiao, Y., Yue, N.: Parameter estimation for nonlinear dynamical adjustment models. Mathematical and Computer Modelling 54(5-6), 1561–1568 (2011) (*cit.* p. 56)

[158] Zhang, Y.K., Bai, E.W., Libra, R., Rowden, R., Liu, H.: Simulation of spring discharge from a limestone aquifer in Iowa, USA. Hydrogeology Journal 4(4), 41–54 (1996) (*cit.* p. 113)

[159] Zhao, Y., Wang, L.Y., Yin, G.G., Zhang, J.F.: Identification of Wiener systems with binary-valued output observations. Automatica 43(10), 1752–1765 (2007) (*cit.* p. 87)

[160] Zi-Qiang, L.: Controller design oriented model identification method for Hammerstein system. Automatica 29(3), 767–771 (1993) (*cit.* p. 14)

Index